Skill-Biased Technological Change
Evidence from a Firm-Level Survey

Donald S. Siegel

1999

W.E. Upjohn Institute for Employment Research
Kalamazoo, Michigan

Library of Congress Cataloging-in-Publication Data

Siegel, Donald S., 1959-
   Skill-biased technological change : evidence from a firm-level survey / Donald S.
Siegel.
    p. cm.
   Includes bibliographical references and index.
   ISBN 0-88099-197-6 (pa. : alk. paper) — ISBN 0-88099-198-4 (cl. : alk. paper)
   1. Labor supply—Effect of technological innovations on—Long Island (N.Y.)—
Surveys. 2. Employees—Effect of technological innovations on—Long Island
(N.Y.)—Surveys. 3. Skilled labor—Long Island (N.Y.)—Surveys. I. Title.

HD6331.2.U52 L667 1999
331.1'09747'21—dc21

                                           99-045380

The facts presented in this study and the observations and viewpoints expressed are
the sole responsibility of the authors. They do not necessarily represent positions of
the W. E. Upjohn Institute for Employment Research.

Cover designed by J.R. Underhill.
Index prepared by Diane Worden.
Printed in the United States of America.

Contents

List of Tables

Acknowledgments

This monograph has benefited from the advice and assistance of many individuals. As a graduate student at Columbia and as a postdoctoral fellow at the National Bureau of Economic Research, I had the good fortune to learn from two of the world's outstanding figures in the economics of technological change, Frank Lichtenberg and Zvi Griliches. Frank stimulated my interest in this topic, while Zvi provided invaluable guidance on the qualitative component of this study (along with several somewhat inscrutable Yiddish proverbs). Frank and Zvi share a common view that economic research is not just a mere academic exercise, but rather a means to help us understand, and ultimately improve, real-world situations. The intellectual influence of these men pervades this monograph.

Numerous economists and experts in human resource management, operations management, and organizational behavior provided useful comments on various aspects of this research. Thanks are due to Mark Doms, Mohan Gopalakrishnan, John Haltiwanger, Boyan Jovanovic, Elliot Kline, Manuel London, Glenn MacDonald, Abagail McWilliams, Catherine Morrison, Susan Slotnick, David D. Van Fleet, David Waldman, Gerrit Wolf, and William Youngdahl; to an anonymous reviewer; and to seminar participants at SUNY-Stony Brook, Arizona State University, the 1994 NBER Summer Institute, the 1995 C.V. Starr Conference on Technologies and Skills at NYU, the 1995 National Academy of Sciences Conference on the Effects of Advanced Manufacturing Technologies and Innovation on Firm Performance and Employment, and the 1996 Academy of Management Meetings. I am especially indebted to Eli Berman for his careful and insightful review of an earlier draft of this monograph. Special thanks to my former colleague at SUNY-Stony Brook, Matthew Sobel, who graciously provided me with the survey data.

I also gratefully acknowledge the financial support of the W.E. Upjohn Institute for Employment Research and SUNY-Stony Brook, which allowed me to deploy an outstanding team of graduate and undergraduate students as research assistants. I am grateful for the diligent assistance of the following students: Sylvia Burns, Nicole Jordan, Hussein Kureshi, Dhruti Pandit, and especially, Bernard Maineville. I am also deeply indebted to the many executives and workers who allowed me to interview them. David Nadziejka, the editor at the Upjohn Institute, has clearly missed his calling in life. Anyone who reads a manuscript with such care and precision should clearly be a Talmudic scholar.

Last, but most importantly, I thank my loving wife, Sandra, and my two wonderful sons, William and Joshua, for providing the warmth and emotional sustenance that made this monograph possible. I dedicate this book to them.

The Author

Donald S. Siegel is Professor of Industrial/Managerial Economics at the University of Nottingham Business School. He received his bachelors, masters, and doctoral degrees from Columbia University. In 1988–1989, Dr. Siegel was a Sloan Foundation Post-Doctoral Fellow in the Productivity and Technical Change Studies program of the National Bureau of Economic Research. He has taught at the State University of New York at Stony Brook (1989–1994) and Arizona State University West (1994–1999), and he was an American Statistical Association/National Science Foundation Senior Research Fellow at the U.S. Bureau of Labor Statistics and a Faculty Research Fellow of the NBER. His primary research interests are productivity analysis and the economic and managerial implications of technological change. Dr. Siegel's research has appeared or is forthcoming in such leading journals in economics and management as the *American Economic Review, Brookings Papers on Economic Activity, Journal of Law and Economics, Journal of Financial Economics, Review of Economics and Statistics, Economic Inquiry, American Journal of Agricultural Economics, Southern Economic Journal, Academy of Management Journal, Strategic Management Journal, IEEE Transactions on Engineering Management,* and *Journal of Management.* He has received grants from the National Science Foundation, the American Statistical Association, the Alfred P. Sloan Foundation, the W.E. Upjohn Institute for Employment Research, and the U.S. Department of Labor.

1 Introduction

THE IMPORTANCE OF THE PROBLEM

For centuries, economists have considered the effects of technological change on labor composition and wages. Malthus (1978), Marx (1967), and Ricardo (1995) all expressed concern about the effects of innovation, especially in the form of new machinery, on the displacement of labor. Joseph Schumpeter (1975) hypothesized that technological change is a force of "creative destruction," which generates new jobs and industries as it destroys existing ones.

In recent years, concerns about the effects of technology on the labor force have been heightened by large-scale corporate downsizing programs and increases in wage inequality. Because these trends have coincided with a large increase in investment in computers, several authors have attributed them, at least in part, to skill-biased technological change; i.e., change that is "biased" by favoring workers with higher levels of education and skill over those with lower levels. This bias occurs because the introduction of a new technology will increase the demand for workers whose skills and knowledge complement that technology.

Many technical advances are labor-saving innovations, enabling companies to eliminate low-skilled positions. This should lead to a shift in labor composition in favor of more highly educated workers. Furthermore, technology may increase the wage premium associated with additional investment in education or skill acquisition. Indeed, as noted by Berman, Bound, and Machin (1998), numerous studies have attributed both the greater wage premium for skill and recent increases in unemployment in Organization for Economic Cooperation and Development (OECD) countries to skill-biased technological change.

Skill-biased technological change has important implications for workers, employers, and public policy. One important issue is whether the implementation of a new technology is accompanied by elements of employee "empowerment" and development strategies. These elements of empowerment could include such factors as additional company-sponsored training, changing job responsibilities, the creation of new jobs and career opportunities, and an increase in employee control, or "voice." It is important to note that advanced manufacturing technologies (AMTs) all involve some transformation of the work environment, because they are integrative (across functional areas of the firm such as manufacturing, marketing, and R&D) and information-intensive, requiring the use of computers. An examination of changes in human resource management (HRM) practices could have important implications for assessing the overall impact of investments in new technology on economic performance, given that recent studies (Bartel 1995; Black and Lynch 1997; Helper 1999) have documented a positive relationship between proxies for worker empowerment (sometimes referred to as "employee involvement" or "voice" practices) and productivity. To the best of my knowledge, however, there is no direct empirical evidence connecting empowerment strategies with specific technological innovations.

Most existing studies of skill-biased technological change have implicitly been based on the concept that technological improvements are homogeneous. In contrast, I examine the labor market consequences associated with different <u>classes</u> of technologies. A disaggregated analysis also provides a more realistic and accurate documentation of changes in HRM policies, such as downsizing and employee empowerment, which emerge after technological change. Specifically, I analyze whether the signs and magnitudes of the skill-bias and employee empowerment effects depend on the type of technology that is implemented. This evidence could be useful to managers, who formulate HRM policies and strategies, and to policymakers, to help target subsidies for training programs and retraining of displaced workers more effectively.

These findings could also have important implications for studies of returns on investment in human capital. Existing theories of human capital imply that under conditions of rapid technological change, creating an environment that fosters organizational learning can increase firm profitability. Such an environment may begin with more adept employees. To build this environment, it is critical for workers to upgrade their skills through training and education, in order to increase proficiency and familiarity with new methods of production. Firms usually fund training for workers who remain with the company in the aftermath of technological changes. Indeed, it appears that this is a rational strategy, given recent evidence indicating positive returns to private sector training (Bartel 1994, 1995; Ichniowski, Shaw, and Prennushi 1997).

Some private sector training is subsidized by the federal government through the Job Training Partnership Act (JTPA). JTPA was established to replace the Comprehensive Employment Training Act (CETA), which targeted job-training programs to the public and non-profit sectors. CETA was terminated in 1982, amid charges of corruption, patronage, and a general sense that the positions created through the program were "make work" jobs. JTPA has a much stronger private sector orientation. It engages unions and firms as partners in training and job-search programs and allocates its funds through state governments, which then distribute the money through local programs. For example, JTPA provided $200 million to the United Automobile Workers to help employees adjust to new "high-performance work organizations," which often accompanied implementation of the new manufacturing technologies (Applebaum and Batt 1994).

Skill-upgrading and employee empowerment are not of much use to workers who lose their jobs. JTPA funds have also been used to assist workers who are displaced in the aftermath of technological change. For instance, in 1993, the Harriman School for Management and Policy at the State University of New York at Stony Brook received JTPA funds, through the Suffolk County Department of Labor, to establish a semester-long "Jobs Project," or Dislocated Worker Training Program, in technology management for 72 older engineers who had been terminated by firms in the local region (Wolf et al. 1995). Although some economists are skeptical about JTPA programs, claiming that they focus on quick solutions and subsidize the most favorable candidates for job placement, bipartisan support for JTPA remains fairly strong. It appears that government funding of formal training programs, administered through local agencies and implemented in community colleges and universities, will continue. The empirical evidence presented in this monograph should be useful to policymakers with such an agenda, by helping them target resources invested in these programs more effectively.

BENEFITS OF EXAMINING THE LONG ISLAND SURVEY

Numerous empirical studies have been conducted on skill-biased technological change. Many of these studies suffered from important limitations. First, most have been based on industry-level data. Firm-level data may be more appropriate because there could be substantial heterogeneity in technology usage and compositional effects within industries. A second limitation is the use of proxies (such as expenditures on R&D and computers) for measuring technological change. One problem with the use of such proxies is that they constitute R&D inputs, rather than outputs (such as patents or the actual implementation of a new production process). Thus, the use of "indicators" limits the accuracy of technology measurement and precludes a precise analysis of timing effects.[1] A third problem is a lack of detailed information on labor force composition. Most datasets identify only two types of laborers: production and nonproduction workers.[2] The underlying problem is that firms are reluctant to provide detailed information on technology usage and workforce characteristics.

The purpose of this monograph is to address the effects of technological change using a new, rich source of firm-level data on technology usage and labor force composition. The empirical investigation is based on a comprehensive, firm-level survey of computer-integrated manufacturing systems (CIMS) usage among Long Island manufacturers. The survey was conducted by a group of professors at the State University of New York at Stony Brook under the direction of Professor Matthew Sobel, with financial support from the New York State Urban Development Corporation.[3]

The primary purpose of the survey was to document the extent of investment in CIMS, which are technologies that use computers to coordinate workers and machines across functional activities, such as production scheduling, procurement, product design, marketing, and distribution, and to identify any obstacles to additional investment. In this monograph, I call these technologies *advanced manufacturing technologies* (AMTs), which is the more commonly used term. This survey provides an ideal data set for exploring the antecedents and consequences of technology adoption, because it contains information on specific types of technologies, the year of implementation, detailed information on labor force composition before and after implementation, and relevant characteristics such as the age of the firm and its R&D expenditures.

It is important to note that technology is not a vague term in this study. In contrast to most existing studies of skill-biased technological change, I directly examine the labor market outcomes associated with the implementation of new manufacturing technologies. Specifically, I examine a well-defined set of 12 AMTs that firms have actually implemented on the factory floor. The companies reported the year of implementation, so I can construct pre- and post-adoption measures of labor composition and relative compensation. AMTs include a wide range of labor saving and quality-enhancing innovations, such as computer-aided design/computer-aided manufacturing (CAD/CAM) systems, computer numerically controlled (CNC) machines, just-in-time (JIT) inventory systems, flexible manufacturing systems (FMS), and robotics (ROB), which allow firms to design, produce, and market new products more effectively and improve manufacturing efficiency. We also have comprehensive information on human resource management strategies that accompany AMT adoption. These data constitute a rich source of information for examining the managerial and policy implications of skill-biased technological change. Based on the survey data, I construct a complete historical profile of each firm's AMT usage and examine the resulting changes in labor force composition and relative compensation over a four-year period. These data enable me to include controls in the econometric model for the endogeneity of technology adoptions, whereas previous studies have generally assumed that technological change is exogenous. I view the implementation of a new technology as a two-stage process. In the first stage, the firm makes a decision to adopt a new advanced manufacturing technology. This leads to an adjustment of the labor force in the second stage.

I also explore whether changes in human resource management policies that enhance employee empowerment arise in the aftermath of technological change. This is crucial because these technologies not only affect labor composition, but also change the work environment for employees in all areas of the firm (i.e., manufacturing, engineering, product development, marketing, R&D, and administrative units). In part, this is because the technological changes promote integration of these functional activities.

Although the survey is quite comprehensive, it cannot completely capture the organization-wide impact of workplace changes that result from the implementation of new technologies. Thus, I also present four case studies of firms that completed the survey, based on 20 plant visits and interviews with company officials and workers. These firms reflect a diverse set of industries and varied experiences with AMT implementation. The case studies highlight some barriers to additional investment in AMT, which suggest some policy responses that might enable firms to surmount these barriers.

Finally, I conduct a disaggregated analysis across two broad classes of AMTs, linked and integrated. This is important because existing studies of skill-biased technological change do not explicitly consider the economic implications of heterogeneous technologies.[4] Specifically, I hypothesize that it is important to distinguish between linked AMTs and integrated AMTs. Linked technologies generally constitute the first generation (or phase) of AMT. Typically, they involve the informational linking of the design and manufacturing functions and establishment of quality and production control practices. Computer-aided design (CAD) is a widely used linked AMT. CAD eliminates most of the drudgery associated with engineering design work, enabling engineers to devote more attention to the creative and evaluative aspects of design. CAD is often "linked" with computer-aided manufacturing (CAM) or with computer-aided engineering (CAE).

Integrated technologies can be thought of as the second generation (or phase) of AMT, involving the integration of other vital components of the manufacturing enterprise such as the material handling and control system. While most linked AMTs are designed to enhance product quality and reliability, the chief purpose of integrated AMTs is to streamline efficiency. Specifically, integrated technologies remove obstacles between physical and organizational entities, reduce costs, and improve flexibility and responsiveness to customers and suppliers. A flexible manufacturing system (FMS) is an example of an integrated AMT that allows the user to respond rapidly to changes in product design and production needs, improves the utilization of machinery and floor space, and reduces work-in-process inventories.

A recent review article (Fine 1993) in the operations management literature theorizes that shifts in the workforce and relative compensation in favor of highly educated workers will be more pronounced for integrated than for linked AMTs. Also, human resource management studies suggest that the effects on certain aspects of employee empowerment will be different for linked and integrated technologies (Appelbaum and Batt 1994; Batt and Appelbaum 1995). To verify these hypotheses and thus, the importance of this distinction, I separately examine the employment and empowerment effects of technology adoption for these two groups of AMTs.

OVERVIEW AND MAJOR CONCLUSIONS

The remainder of this monograph is organized as follows. Chapter 2 presents a comprehensive review of the recent literature on the employment and wage effects of technological change. Both labor and productivity economists have addressed this subject using different methodological approaches and a wide range of datasets.

Chapter 3 contains an extensive description of the database and the survey methodology. First, the survey design and some relevant aspects of the business environment on Long Island are discussed. I present summary statistics for the Long Island sample and a discussion of the representativeness of the sample. Although the study focuses on one particular region, the findings have important implications for national technology policy. The chapter also discusses the two main hypotheses of the monograph. The first is the "non-neutrality" of technological change with respect to the composition of the labor force, and the second is the importance of organizational learning in the process of adopting a new technology. I also discuss some econometric issues relating to assessing the impact of technological change on labor composition and show how I address these issues in the empirical estimation.

Chapter 4 provides an in-depth explanation of the salient characteristics of the advanced manufacturing technologies; examples are provided for each AMT. I postulate that it is important to distinguish between linked and integrated AMTs because linked and integrated AMTs could have differential impacts on labor composition and other aspects of the work environment. One characteristic of the work environment that could change in the aftermath of technological change is the level of employee empowerment. In the final section of this chapter, I provide an operational definition of empowerment and hypothesize that AMT investment, especially linked AMTs, will lead to greater employee empowerment.

Chapter 5 begins with a summary of the set of hypotheses I have outlined in previous chapters and wish to test empirically. The chapter continues with the empirical results regarding the determinants and labor market outcomes of technology adoption, including its impact on employment, labor composition, and proxies for employee empowerment, and how these effects differ for linked and integrated AMTs.

In Chapter 6, I present the case studies of four Long Island manufacturers and evidence from our visits to 16 additional firms in the sample. This qualitative evidence elaborates on certain points that could not be addressed in the survey and the subsequent statistical analysis.

Conclusions and policy implications are discussed in Chapter 7. The following is a summary of the key findings:

1. Technological change is associated with downsizing and a shift in labor composition in favor of workers with higher levels of education.

2. The probability of technology adoption is uncorrelated with the age of the firm but is positively associated with firm size, R&D intensity, and previous technology adoptions.

3. A factor analysis confirms the validity of the distinction between linked and integrated AMTs.

4. Recomposition in favor of more highly educated workers appears to be most strongly associated with integrated AMTs. It is important to be mindful of these differential impacts when formulating technology policies.

5. New technologies lead to greater empowerment for workers, where empowerment is defined as training of existing personnel, changing job responsibilities, creating new jobs and career opportunities, and increasing the extent of employee control. I find that empowerment is more closely associated with linked, rather than integrated, AMTs.

6. The field interviews appear to confirm the statistical findings. In the firms I examined, AMTs (especially, integrated AMTs) were indeed associated with personnel reductions and skill up-grading.

7. The field interviews also revealed two major obstacles to additional investment in new technology: difficulties in quantifying the benefits from technological investments and the high cost of customizing software to fit company needs.

Notes

1. An exception is a paper by Doms, Dunne, and Troske (1997), which explores timing issues based on confidential, plant-level, U.S. Census data with direct measures of technological change.
2. A notable exception is Lynch and Osterman (1989), who examined compositional effects of technological change for 10 occupational classes of workers.
3. The CIMS project was sponsored by the New York State Urban Development Corporation under the auspices of the Long Island Office of the New York State Department of Economic Development. I am deeply indebted to Professors Gerrit Wolf and Manny London, and especially to Professor Matthew Sobel, for providing me with these data.
4. Contrast this to the literature on the impact of new technology on total factor productivity, where it is common to conduct such a disaggregated analysis. As discussed in Lichtenberg and Siegel (1991), researchers have reported the "returns" on various types of R&D investments, such as product vs. process innovation, basic research vs. applied R&D, or privately funded vs. publicly funded R&D.

2 Previous Studies of Skill-Biased Technological Change

In recent years, there has been a widening of the wage differential between low-skilled and high-skilled workers (Murphy and Welch 1992; Bound and Johnson 1992). This has occurred despite a large increase in the number of high-skilled workers. One explanation for this increase in the rate of return on investment in education is "skill-biased" technological change.[1] This hypothesis, advanced by Nelson and Phelps (1966), Griliches (1969, 1970), and Welch (1970), maintains that the value of education is enhanced by technological change because greater knowledge or skill enables companies to implement new technologies more effectively. Bartel and Lichtenberg (1987) modify this theory by asserting that the comparative advantage of highly skilled or highly educated workers in implementing new technologies arises from their ability to solve problems and adapt to change in the work environment. These models predict that technological change is biased (non-neutral) with respect to labor, having disproportionate effects on different classes of workers.

Determining whether technological change is non-neutral is important for two reasons. First, conventional measures of productivity growth are estimated under an assumption of neutrality; imposing non-neutrality when it is unwarranted could lead to biased and imprecise measures of productivity.[2] Griliches (1996) notes that the earliest studies of the role of education in production were concerned with accounting for the large productivity "residual," the portion of economic growth that cannot be explained by measured capital and labor inputs. The concern is that unmeasured improvements in the quality of labor could lead to underestimation of real labor input and, thus, overestimation of the true rate of growth in total factor productivity. In this regard, studies of the sources of economic growth ("growth accounting" studies) have reported quality-adjusted measures of labor input based on indexes of workforce educational attainment (see Denison 1962; Jorgenson and Griliches 1967; Jorgenson, Gollop, and Fraumeni 1987; and Dean, Kunze, and Rosenblum 1988).

Second, assessing the non-neutrality of technological change is important to our understanding of several important trends in the labor market. Accordingly, labor economists tend to focus on the wage and employment implications of skill-biased technical change (e.g., Mincer 1989; Davis and Haltiwanger 1991; Levy and Murnane 1992; Katz and Murphy 1992; Murphy and Welch 1992; Juhn, Murphy, and Pierce 1993; Goldin and Katz 1996; Autor, Katz, and Krueger 1997; Bartel and Sicherman 1999; Haskel 1999; Haskel and Heden 1999). Such studies attempt to determine how much of the rise in wage inequality and the concomitant increase in the demand for highly skilled and highly educated workers can be attributed to the use of new technologies. Davis and Haltiwanger (1991) documented a large increase during the 1980s in the earnings differential between nonproduction and production worker wages. The authors attributed these changes to non-neutral technological change, which has increased the relative earnings of highly skilled workers.[3]

Consequently, economists in two fields, productivity analysis and labor, have generated a large body of empirical evidence of the complementarity between highly skilled or educated labor and technical capital (Table 2.1).[4] Note that despite the different methodologies and analysis of data from many countries at different levels of aggregation (individual, plant, firm, and industry levels), each study provides evidence that is consistent with some aspect of the theory of skill-biased technological change; that is, these researchers generally find that some proxy for technological change (R&D, computers, adoption of advanced manufacturing technologies) is positively correlated with wages and shifts in labor composition in favor of highly skilled or highly educated workers.

Reflecting the differences in perspectives, basically two types of studies are represented in Table 2.1. The first type uses a production or cost function framework, usually based on estimation of a reduced-form model. Many of these studies report the results of industry-level regressions of changes in employment shares or wages on proxies for technological change, such as R&D investment.

Berman, Bound, and Griliches (1994) found a positive association between investments in computers and R&D and changes in nonproduction workers' share of the industry wage bill in 450 U.S. manufacturing industries. The latter is interpreted as indicative of "skill upgrading." The authors also report that much of this skill-upgrading occurred within industries, which implies that skill-upgrading is not the result of shifts in product demand. Using similar methods, Mishel and Bernstein (1994) included the employment share of scientists and engineers in the industry as an additional indicator of technological change. They reported a similar positive correlation between proxies for technological change and shifts in demand in favor of highly educated workers. However, they did not find that this relationship had become stronger in the 1980s.

Other industry-level studies are consistent with these results. Berndt, Morrison, and Rosenblum (1992) reported a positive correlation between high-tech office equipment and the demand for white collar workers (2-digit SIC level). In a recent study (Siegel 1997) based on estimation of a latent variables model, I found a positive association between proxies for labor quality and computer investment for 293 U.S. manufacturing industries (4-digit SIC level). Finally, Bartel and Lichtenberg (1987, 1990) reported that the demand for highly educated workers (as proxied by the wage rate) is inversely related to the age of an industry's technology (approximately 3-digit SIC level). This is consistent with the authors' theoretical model, which asserts that the demand for "learning" is highest when a firm implements a new technology. Following the rationale of "efficiency wage" models, they argued that a higher wage is also needed to elicit higher levels of effort in the aftermath of technological change.

An analysis of industry-level data from foreign countries yields similar patterns. Berman, Bound, and Machin (1998) determined that changes in the employment structure in favor of highly educated workers are evident in nine OECD countries. The authors concluded that these wage and employment shifts can be linked to technological change. Also, the magnitudes of these linkages are quite similar across countries. Further international evidence is provided by Park (1996), who reported a positive correlation between labor productivity growth and the proportion of multiskilled workers in Korean manufacturing industries. For two-digit manufacturing industries in Canada, Betts (1997) estimated a fully specified translog cost function model with time, a proxy for technical change, as an additional argument. He found evidence of non-neutral technical change away from blue-collar labor in 16 out of 18 industries. In the United Kingdom, Haskel

Table 2.1 Twenty-Seven Recent Empirical Studies of Skill-Biased Technological Change

Authors	Methodology	Country	Level of aggregation	Indicators of technical change	Measures of labor input	Results
Bartel & Sicherman (1999)	Estimation of wage equations	U.S.A.	Worker data (NLSY) matched to industry-level data	TFP growth, patents, scientists & engineers, expenditures on computers & R&D	Nonproduction and production workers	Positive correlation between wages and proxies for technical change, which is stronger for nonproduction workers than for production workers; the wage premium is attributed to the greater demand for ability in industries experiencing technical change
Haskel (1999)	Regressions of changes in the relative wages of skilled and unskilled workers on computers	United Kingdom	3-digit SIC industry	Dummy variable denoting whether a plant introduced new equipment based on microchip technology	Skilled and unskilled workers	Positive correlation between relative wages and computers; wage premium for skill rose by 13% in the 1980s in the U.K.; computers account for about half of this increase
Haskel & Heden (1999)	Regressions of changes in wage bill share for four classes of workers on computers and R&D	United Kingdom	Plant and industry level	Expenditures on computers, R&D; proportion of firms in sector using computers in production process	Nonmanual and manual workers split into skilled and unskilled categories	Positive correlation between the relative wages of skilled nonmanual workers and computers (also R&D); computerization reduces the demand for manual workers (both skilled and unskilled workers)
Berman, Bound, & Machin (1998)	Cross-country correlations of within-industry changes in the proportion of nonproduction workers	9 OECD countries	2- and 3-digit SIC industries	Expenditures on computers, R&D	Employment and wage shares for production & nonproduction workers	Positive correlation across 9 OECD countries in within-industry changes in shares of nonproduction workers

Study	Method	Country	Data	Technology measure	Labor composition	Findings
Hildreth (1998)	Estimation of wage equations	United Kingdom	Data on workers and plants that employ them	Dummy variable denoting whether a plant invested in a new product or process technology	No data on labor composition	Workers employed at plants that invest in new process technologies earn higher wages; rent-sharing (between workers and firms) is strongest in high-tech plants
Autor, Katz, & Krueger (1997)	Estimation of wage equations	U.S.A.	Worker data (CPS)	Dummy variable for whether worker uses a computer	Detailed data on workers: age, sex, race, union status, region	Workers who use a computer at work earn a 17–20% wage premium; computers could account for 30–50% of the recent increase in demand for highly skilled workers
Betts (1997)	Estimation of translog cost function	Canada	2-digit SIC industry	Time	Nonproduction and production workers	Strong evidence of biased technical change away from blue-collar workers
Dinardo & Pischke (1997)	Estimation of wage equations	Germany	Worker data qualification and career survey	Dummies for whether worker sits down, uses a telephone, calculator, pen and pencil	Detailed data on workers: age, sex, union status	Workers who use a computer earn a wage premium, but so do those who sit down while they work or use a calculator, telephone, or pen and pencil
Doms, Dunne, & Troske (1997)	Estimation of wage equations	U.S.A.	Data on workers and plants that employ them	Establishment-level data on AMTs	Occupational mix, education	Positive correlation between technology usage and levels of (but not changes in) wages, skill, and education
Dunne, Haltiwanger, & Troske (1996)	Regression of changes in nonproduction workers' share in employment on R&D and number of AMTs adopted	U.S.A.	Plant level	Firm-level measures of R&D; plant-level measures of AMTs	Nonproduction and production workers	Positive correlation between changes in nonproduction labor share and R&D, but not between changes in nonproduction labor share and AMT usage

Authors	Methodology	Country	Level of aggregation	Indicators of technical change	Measures of labor input	Results
Siegel (1997)	Regressions of an index of labor quality on measures of the rate of investment in computers	U.S.A.	4-digit SIC industry	Expenditures on computers, R&D	Age, education cells for nonproduction and production workers	Positive correlation between indicators of labor quality and investment in computers
Machin (1996)	Regression of changes in skilled workers' share in employment on various proxies for technological change	United Kingdom	Industry and plant level	R&D intensity, innovation counts, introduction of microcomputers	Industry level: manual and nonmanual workers; plant level: exmployment shares for 6 skill groups	Positive correlation between changes in nonmanual labor share and R&D & innovations; skill upgrading associated with computers only for workers with highest level of skill
Park (1996)	Regressions of proportion of multiskilled workers on labor productivity growth and capital-labor ratio	South Korea	2-digit SIC industry	Growth in labor productivity	All workers, excluding unskilled	Positive correlation between labor productivity growth and the proportion of multiskilled workers in Korean manufacturing
Van Reenen (1996)	Estimation of wage equations with panel data on innovations	United Kingdom	Firm level	Number of firm innovations and patents granted	No data on labor composition	Innovative firms pay above-average wages
Chennells & Van Reenen (1995)	Estimation of wage equations	United Kingdom	Plant level	Plant-level data on technology usage	3 classes of workers: skilled, semi-skilled, and unskilled	Positive correlation between technology usage and wages

Study	Method	Country	Level	Technology measure	Labor classification	Findings
Dunne & Schmitz (1995)	Regressions of change in nonproduction workers' share in employment on number of AMTs adopted	U.S.A.	Plant level	Establishment-level data on AMTs	Nonproduction and production workers	Plants with a high rate of technology adoption pay higher wages and employ a larger percentage of nonproduction workers
Entorf & Kramarz (1995)	Estimation of wage equations	France	Data on workers and firms that employ them	Firm-level data on usage of 3 computer-based technologies	Occupational mix: unskilled and skilled blue-collar, clerks, managers, engineers, professionals	Positive correlation between technology usage and wages; highest wage premiums earned by those with the lowest level of skill
Regev (1995)	Estimation of production function	Israel	Firm level	Technology index based on quality of labor and capital and R&D investment	No decomposition of labor	Technology-intensive firms pay higher average wages, generated new jobs during a period of downsizing
Reilly (1995)	Estimation of wage equations	Canada	Data on workers and plants that employ them	Dummy variable denoting whether firm has access to computers	Detailed data on workers: occupation, industry, age, tenure, region, experience	Workers that have access to computers earn a 13% wage premium
Berman, Bound, & Griliches (1994)	Regressions of changes in nonproduction workers' share in total wages on computers and R&D	U.S.A.	4-digit SIC industry	Expenditures on computers, R&D	Employment and wage shares for production and nonproduction workers	Positive correlation between computers and R&D and changes in nonproduction workers' share in employment and wages

Authors	Methodology	Country	Level of aggregation	Indicators of technical change	Measures of labor input	Results
Mishel & Bernstein (1994)	Regression of changes in employment shares for 5 educational classes of workers on technology proxies	U.S.A.	2-digit SIC industry	Computer and equipment capital per worker, employment share of scientists and engineers	Education, gross and residual wage inequality	Positive association between technology proxies and proportion of educated labor, but no stronger in the 1980s than in the 1970s
Krueger (1993)	Estimation of wage equations	U.S.A.	Worker, CPS	Dummy for whether worker uses computer	Detailed data on workers: 8 occupations, age, sex, union status	Workers who use a computer at work earn a 10–15% wage premium
Berndt, Morrison, & Rosenblum (1992)	Regressions of labor intensity measures on "high tech office equipment," capital intensity	U.S.A.	2-digit SIC industry	"High-tech office equipment," capital stock	Age, education for production and nonproduction workers	Positive correlation between share of "high-tech office equipment" and employment share of nonproduction workers
Bartel & Lichtenberg (1990)	Estimation of wage equation	U.S.A.	Workers in 3-digit SIC industry	Proxies for the age of the capital stock, R&D	Age, education, sex	Inverse relationship between the age of technology and wages of highly educated workers
Lynch & Osterman (1989)	Estimation of labor demand curves for 10 occupational classes of workers	U.S.A.	Firm level (n = 1), telephone company	Technical change in switching equipment, production of operators, capital expenditures	10 occupational classes of workers	Innovation favors professional employees, also leads to greater centralization
Bartel & Lichtenberg (1987)	Estimation of restricted (variable) labor cost function	U.S.A.	3-digit SIC industry	Proxies for the age of the capital stock	Age, education, sex	Inverse relationship between technology and the percentage of labor cost devoted to highly educated workers

| Osterman (1986) | Regression of changes in employment after the installation of computers | U.S.A. | 2-digit SIC industry | Industry measure of total computer memory | Several occupational classes: clerks, non-data-entry clerks, and managers | Computerization reduces employment of clerks and managers, not as much for managers in the long run |

(1999) and Haskel and Heden (1999) report that computerization increases the demand for skilled workers in the manufacturing sector.

The second type of study follows the standard approach in labor economics: the estimation of wage equations. One of the first studies to link changes in the wage structure at the micro level to technology use was by Krueger (1993), who used data from the Current Population Survey (CPS). This survey contained a question on whether an employee uses a computer at work. Krueger reported that workers who use a computer on the job earn a wage premium (10–15%) relative to observationally equivalent workers in the October 1984 and 1989 waves of the CPS. Reilly (1995) found that Canadian workers with access to computers earned a 13% wage premium during an earlier period.

In a study titled "Computing Inequality: Have Computers Changed the Labor Market?", Autor, Katz, and Krueger (1997) updated Krueger's 1993 study, showing that the wage premium had increased in the last decade to approximately 17%.[5] More importantly, the authors concluded that investments in computers could account for as much as 35–50% of the increase in the growth in demand for highly skilled workers.

Because industry-level studies could be subject to aggregation biases, it is more desirable to examine the impact of technology on wages and labor composition at the plant or firm level, because there could be substantial variation in these effects within industries. There have also been several firm- and plant-level studies of skill-biased technical change in the U.S.A., France, and the United Kingdom.

The first firm-level study was conducted by Lynch and Osterman (1989), who estimated labor demand curves for workers employed by a single firm in the telecommunications industry. The authors reported that technological innovations stimulated an increase in the demand for technical and professional workers.

A cross-sectional, plant-level study by Dunne and Schmitz (1995) was based on the U.S. Survey of Manufacturing Technology (SMT).[6] This file contains detailed information on adoptions of advanced manufacturing technologies by thousands of plants in five 2-digit SIC industries (SICs 34–38), average wages, and limited information on labor composition (production versus nonproduction workers). The authors reported that technology-intensive plants pay higher wages than less-technology-intensive plants within the same industry.

Evidence from labor markets in the United Kingdom and Israel is consistent with this finding. Van Reenen (1996) examined panel data on wages and innovation for a sample of British firms whose shares were publicly traded for at least five years between 1976 and 1982. The data on innovations were derived from the Science Policy Research Unit (SPRU) database, which contains detailed information on successful commercial innovations in Great Britain between 1945 and 1983. Using both static and dynamic instrumental variables (generalized methods of moments, or GMM) estimation to control for the endogeneity of innovations, he concluded that innovative firms pay above-average wages.[7] Regev (1995) estimated a simple production function model for a panel dataset of 2500 Israeli firms. He constructed a "technology index" for each firm, consisting of measures of the quality of labor, capital, and R&D investment. He reported that technology-intensive firms pay above-average wages and are consistently more productive than other firms in the same industry. Regev also found that these firms demonstrated net job creation during a period when many companies were downsizing.

One of the most important developments in empirical analysis of skill-biased technological change has been the creation of databases that match workers to their place of employment. Traditional studies of the labor supply behavior of individuals have suffered from limited information regarding the demand for a worker's labor. To understand the nature of this demand and to help sort out the determinants of intra- and inter-industry wage differentials, it is helpful to simultaneously explore data on the characteristics of workers and firms. Note that conventional datasets used in labor market studies, such as the CPS, the National Longitudinal Survey (NLS), or the Panel Study of Income Dynamics (PSID), do not contain detailed information on the employer.

Researchers at the U.S. Census Bureau (see Troske 1994) have constructed the Worker-Establishment Characteristic Database (WECD), a file that links detailed demographic data from the 1990 Decennial Census to comprehensive information on plants contained in the Longitudinal Research Database (LRD). The LRD is a compilation of data on establishments from the Census of Manufactures and the Annual Survey of Manufacturers. This file has also been linked to the

Survey of Manufacturing Technology, which provides detailed information on advanced manufacturing technology usage. There are now two cross sections of the SMT, a 1988 and a 1993 version.

The linked version of the WECD and SMT has been analyzed by Dunne, Haltiwanger, and Troske (1996) and Doms, Dunne, and Troske (1997). Both studies reported a positive correlation between technology usage and levels of (but not changes in) wages and education. The authors also concluded that high-wage, high-skill plants are more likely to adopt new technologies. They found no evidence of workforce adjustment or "skill-upgrading" in the aftermath of technology adoption. While the cross-sectional analysis of wage and compositional effects is much richer than the previous census study (Dunne and Schmitz 1995), the longitudinal analysis suffers from two important limitations. First, they can only measure changes in employment and wages for two types of employees: production and nonproduction workers. Second, they cannot identify the exact year of technology adoption, which precludes a precise analysis of timing effects.

Matched employee-employer datasets have also been constructed in England and France. Chennells and Van Reenen (1995) examined the 1984 Workplace Industrial Relations Survey (WIRS), a plant-level survey conducted in the United Kingdom. The WIRS survey contained a question that asked managers whether the plant has implemented a new computer technology. The authors reported that, for workers in all four skill categories (skilled, semi-skilled, unskilled, clerical), there was a positive association between wages and technology usage. They found technological wage premiums of about 5% for skilled workers and about 10% for semi-skilled and unskilled workers, with a premium of 7% overall.

Machin (1996) linked the WIRS survey to the SPRU innovation database. This enabled him to construct two additional proxies for technological change: R&D intensity and innovation counts. He then regressed changes in employment shares for six classes of workers on these proxies. R&D and innovations were positively associated with shifts in labor composition in favor of highly educated workers. However, computers were associated with skill upgrading only for workers with the highest level of education or skill.

Entorf and Kramarz (1995) examined a French matched employee-employer panel dataset with detailed measures of labor composition and technology usage. The authors also found a positive correlation between technology usage and wages. Interestingly, they found the highest wage premiums accrued to workers with the lowest level of skill. Their conclusion was that for many highly educated and skilled workers, proficiency with a new technology is expected, and thus is already factored into the current wage.

It is interesting that both Chennells and Van Reenen (1995) and Entorf and Kramarz (1995) concluded that it is unlikely that new technologies "cause" higher wages, casting doubt on the conventional interpretation of the wage premium on computers or new technology as reflecting true "returns" (see DiNardo and Pischke 1997). Of course, it is difficult to sort out these issues without more precise information on the timing of innovations.

A review of the literature provides strong empirical support for the theory of skill-biased technological change. That is, with virtual unanimity, authors found a positive association between some proxy for technological change and changes in labor force composition and relative compensation in favor of highly educated workers. This finding is consistent across different countries, time periods, methodologies, and levels of aggregation.

One important question that has not been addressed in existing studies is whether there is heterogeneity in the employment and wage effects across different classes of technologies. Thus, to the best of my knowledge, I will be the first to examine whether the nature of the skill-bias differs for two types of technologies, linked versus integrated.

Notes

1. Trade is also alleged to have increased the earnings gap. According to this view, cheap imports produced by unskilled workers have reduced the wages of low-skilled U.S. workers. Most of the empirical evidence (see Lawrence and Slaughter 1993) is not consistent with this hypothesis.
2. The usual assumption is that technological change is "disembodied"; i.e., that it affects each factor of production in the same manner.
3. Groshen (1990) finds similar trends in more detailed occupations. She concludes that these changes reflect higher returns to vocational-specific training.
4. These studies have used numerous measures of technological change, such as spending on computers, spending on R&D, and the adoption of advanced manufacturing technologies.

5. A recent paper by DiNardo and Pischke (1997), based on the German equivalent of the CPS and the same research design as the Krueger study, questions this interpretation. Although, like Krueger, these authors also found that workers who use a computer earn higher wages, they also reported wage premiums associated with the use of a calculator, a telephone, and a pen and pencil, as well as with whether a worker is seated while on the job.
6. We will discuss this national survey in greater detail in Chapter 3.
7. Van Reenen (1997) used the SPRU dataset to examine the relationship between employment growth and technological innovation. Contrary to most studies, he found that technological change is associated with higher levels of employment. He did not examine compositional effects.

3 The Survey and the Econometric Model

THE SURVEY METHODOLOGY

For this survey, a potential subject pool was identified by targeting all manufacturing companies on Long Island with at least 100 employees, as well as 49 smaller manufacturers. These firms were identified from the 1988 *Long Island Business Directory* and with databases provided by the Long Island Regional Office of the New York State Department of Economic Development. These sources provided a total of 403 firms. However, some of these companies had to be eliminated because they either

- moved their manufacturing operations off Long Island,
- no longer engaged in manufacturing, or
- went out of business.

This reduced the sample to 369 firms.

The full survey consisted of over six pages of detailed questions on the determinants and outcomes of investment in advanced manufacturing technologies. Firms were asked to report comprehensive information on labor composition, the methods and cost of AMT implementation, and R&D expenditures by type, character of use, and source of funds. Additional descriptive questions were on the expected and actual benefits of investment in new technologies and on changes in the organizational environment that resulted from implementation, including changes in job responsibilities and differences in reporting relationships. We also asked several questions about customers and suppliers and about a wide range of firm characteristics.

For this study, the key data items are

- the extent of AMT implementation;
- the year of AMT implementation;
- the methods of AMT implementation (which include our proxies for employee empowerment);

25

- levels of employment for the years 1987–1990 for six types of workers: managerial and supervisory, technical and professional, scientists and engineers (R&D staff), clerical and administrative, direct (production) labor and supporting personnel, and other production employees (generally, service workers);
- the ratio of R&D expenditures to sales by type, character of use, and source of funds;
- age of the firm; and
- primary industry (3-digit SIC).

Given the complex nature of the survey, it was important to identify the proper company official to complete it. A cover letter requesting cooperation was first sent to the firms' vice presidents of engineering and manufacturing, after which each firm was contacted by phone to determine whether the potential respondent could accurately complete the survey. These phone conversations revealed that for many smaller firms, the CEO or company president was the appropriate respondent.

With this additional information, the survey was mailed to the 369 firms during the fall of 1990. Seventy-nine firms chose to participate with one individual per firm filling out the survey.[1] In several instances, respondents submitted partially completed surveys, and in these cases, a team of five graduate students telephoned the company executives and helped them complete the questionnaire.

The overall response rate was approximately 21%, which is quite high for a complex survey of this nature. This compares favorably with an 18.0% response for a recent firm-level survey of AMT usage by Dean and Snell (1996). It is also significantly higher than the 6.5% response rate reported by Delaney, Ichniowski, and Lewin (1989) in their business-level survey of HRM practices.[2]

One limitation of our survey, relative to several proprietary establishment-level datasets housed at statistical agencies in the U.S.A. and France (Doms, Dunne, and Troske 1997 and Entorf and Kramarz 1995) is that we do not have information on worker characteristics. It would certainly be useful to have information on the educational level, experience, and other relevant human capital characteristics of individual workers. It is important to note that many theoretical models in labor

economics highlight the importance of employee and employer characteristics in labor market outcomes, such as job "matching," turnover,
and investment in human capital. Unfortunately, most empirical tests
of these models have been based on establishment-level datasets with
no employee information or on worker surveys that contain little information on the characteristics of the worker's employer. This could
result in biased and imprecise estimates of these models.

As discussed in Chapter 2, Doms, Dunne, and Troske (1997)
examined an employee-employer matched dataset (a linked version of
the Worker-Establishment Characteristic Database [WECD] and the
National Survey of Manufacturing Technology) and found that firms
with more skilled workforces are more likely to adopt new advanced
manufacturing technologies. Clearly it would be useful to have this
information in the first stage of my econometric analysis, which examines the probability of adoption. However, firms do report the rate of
R&D investment, which has been demonstrated in several studies (Berman, Bound, and Griliches 1994; Bartel and Lichtenberg 1990; Siegel
1997) to be strongly correlated with worker skill and other human capital characteristics. Finally, I believe that the absence of data on individual workers is not a significant drawback for this study, given my
primary focus on the effects of technology adoption on labor composition. Information on individual workers would clearly be useful if the
main goal of this study were to explain changes in individual worker
wages, but that is not its purpose.

SUMMARY STATISTICS FOR
THE LONG ISLAND SAMPLE

Our 79 respondents constituted over 85% of manufacturing
employment on Long Island. Table 3.1 presents an industry distribution for the respondents, showing that approximately one-third of the
firms are in the electronics industry (SIC 36). This is not surprising,
given that the region accounts for approximately 35% of defense
employment in New York State. Furthermore, response rates in the
industries of Table 3.1 appear to be rather high. Table 3.2 shows the
coverage for industries (at the three-digit SIC level) that the Depart-

Table 3.1 Industry Distribution for the 79 Surveyed Companies

SIC	Industry	Number of firms	Percentage	Cumulative number	Cumulative percentage
25	Furniture	1	1.3	1	1.3
26	Paper and allied products	1	1.3	2	2.6
28	Chemicals	8	10.1	10	12.7
30	Rubber and miscellaneous products	3	3.8	13	16.5
32	Stone, clay, and glass products	3	3.8	16	20.3
33	Primary metal industries	1	1.3	17	21.6
34	Fabricated metal products	8	10.1	25	31.7
35	Machinery, except electrical	9	11.4	34	43.1
36	Electric and electronic equipment	25	31.6	59	74.7
37	Transportation equipment	5	6.3	64	81.0
38	Instruments	8	10.1	72	91.1
39	Miscellaneous manufacturing	7	8.9	79	100.0

Table 3.2 Coverage of 11 Defense-Related Industries in the Long Island Survey (1990)

SIC	Industry	Sample employment	Total Long Island employment	Percentage
346	Metal forgings and stampings	650	802	81.0
354	Metalworking machinery	1,509	1,703	88.6
356	General industry machinery	2,101	2,669	78.7
361	Electric distribution equipment	875	1,002	87.3
362	Electrical industrial apparatus	3,895	4,011	97.1
366	Communications equipment	5,978	6,306	94.8
367	Electronic components	9,987	10,231	97.6
372	Aircraft and parts	28,694	29,604	96.9
373	Ship and boat building and repair	291	305	95.4
381	Search and navigation equipment	12,212	15,296	79.8
382	Measuring and controlling devices	11,105	13,012	85.3
Total		77,297	84,941	91.0

ment of Defense has identified as primarily defense-related; in 5 out of 11 of these industries, the survey's coverage (in terms of employment) is over 90%.

Table 3.3 contains descriptive statistics on the age, size, and R&D intensity (R&D/sales ratio) of the 79 firms. The median age and size of the firm are 30 years and 150 employees ($12 million in sales), respectively. The sample appears to be representative along these two dimensions; according to Dunne, Roberts, and Samuelson (1989), the median age and employment of the representative manufacturing firm are 26.8 years and 128 employees, respectively.[3] That the sample is weighted towards high-tech, defense-related firms is reflected in the mean R&D intensity of 5.7%, which exceeds the corresponding figure for representative manufacturing firms of 3.6% (National Science Foundation 1989).

Table 3.4 provides additional evidence on our first concern, the representativeness of the Long Island response group. In this table, the rate of technology adoption in our survey sample is cross-classified with the industry and the size and age of the firm, and is compared with statistics from a recent national survey of advanced manufacturing technology (U.S. Bureau of the Census 1989).[4] The national survey was conducted only in SICs 34–38. Comparing firms in the same industry and of similar age and size, I find that the rate of technology adoption in the Long Island survey sample is roughly comparable to U.S. averages.

To summarize, the sample appears to be representative along the dimensions of age and size, but weighted towards high-tech, defense-related firms. This highlights the importance of including industry controls, because the defense industry experienced substantial downsizing in the late 1980s. These controls are crucial to my objective of isolating and measuring the employment effects of technological change.

A second concern is whether we observe a sufficient number of technology adoption events during the survey "window." As noted earlier, I observed the complete historical profile of the firm's investment in AMTs; however, I only observed the composition and levels of employment over a four-year period, 1987–1990. Consequently, my ability to examine the employment effects of technological change is limited to these four years. Thus, it is critical to identify how many firms adopted new technologies during the mid to late 1980s.

Table 3.3 Characteristics of 79 Long Island Manufacturers in 1990

Characteristic	Mean	Median	Q1	Q3	Minimum	Maximum
Age of firm (years)	32.3	30.0	17.0	40.0	4.0	140.0
Sales ($ millions)	6.1	12.0	5.0	23.0	0.12	2,500.0
Employment	522	150	81	300	5	17,000
R&D/sales (%)	5.7	4.5	2.0	6.0	0.0	71.0

**Table 3.4 Long Island Survey Firms vs. U.S. Rates of Technology
Adoption by Industry, Size, and Age of Firm (% of firms)**

	AMTs used	
Group	0	≥ 1
By industry		
SIC 34		
L.I.	25.0	75.0
U.S.	32.6	67.4
SIC 35		
L.I.	22.0	78.0
U.S.	18.1	81.9
SIC 36		
L.I.	12.5	87.5
U.S.	17.1	73.4
SIC 37		
L.I.	20.0	80.0
U.S.	28.2	71.8
SIC 38		
L.I.	12.5	87.5
U.S.	21.3	78.7
By firm size (number of employees)		
≤99		
L.I.	28.6	71.4
U.S.	32.6	67.4
100–499		
L.I.	11.9	88.1
U.S.	18.1	71.9
>500		
L.I.	12.5	87.1
U.S.	17.1	73.4
By firm age		
≤15 years		
L.I.	29.1	70.9
U.S.	32.6	67.4

Group	AMTs used	
	0	≥ 1
16–30 years		
L.I.	21.4	78.6
U.S.	18.1	71.9
>30 years		
L.I.	13.3	86.7
U.S.	17.1	73.4

SOURCE: Long Island sample, CIMS survey conducted by SUNY-Stony Brook; National Sample, U.S. Bureau of the Census (1989) (national survey conducted only in SICs 34–38).

In Tables 3.5 and 3.6, I present distributions for the number of technology adoptions and the year of implementation. Table 3.5 indicates that a large proportion of companies (approximately 58%) have adopted at least two technologies. Having multiple adopters in the survey is desirable because it allows us to examine the relationship between the number of adoptions and employment and wage effects. It does, unfortunately, make it difficult to separate out the employment and wage effects of implementing a specific technology.

The overwhelming majority of adoptions (82.3%) occurred during the years 1984–1990. Of course, the degrees of freedom are limited by the fact that I don't observe four years of post-event data for each technology adoption; that is, the final sample will not consist of a balanced panel. Still, there appears to be a sufficient number of events for efficient econometric estimation.

It is also important to avoid confusion about when the event actually begins. For this reason, we asked firms to report both the year they decided to adopt the technology and its first year of use. The survey responses underscore the importance of this distinction, because the planning period before implementation can sometimes exceed a year. When a firm decides to adopt a new technology, managers must decide which computer systems and software to purchase, how to customize software, how to retrain workers, and whether consulting services will be needed. This process of gearing up for the new technology can

**Table 3.5 Distributions of Number of AMTs Adopted
by the 79 Survey Firms**

Number of technologies	Number of firms	Percent	Cumulative number	Cumulative percent
0	13	16.5	13	16.5
1	20	25.3	33	41.8
2	12	15.2	45	57.0
3	12	15.2	57	72.2
4	6	7.6	63	79.8
5	4	5.1	67	84.9
6	9	11.5	76	96.4
7	1	1.2	77	97.6
8	1	1.2	78	98.8
9	1	1.2	79	100.0

sometimes be time-consuming. Before implementation, companies are concerned about integrating the new technology into the organization as quickly and smoothly as possible. Any delays could be extremely costly.

For the 20 firms in our survey that I interviewed, final integration of the new technology occurred, on average, about four to six months after the initial stages of implementation.[5] Similarly, Dean (1987) and Foston, Smith, and Au (1991) reported that, for most companies, an AMT is completely integrated into the firm's daily operations about five to six months after the initial implementation. I am reasonably confident that for an overwhelming majority of technology adoptions in our sample, full integration occurs by the end of the first year of use.

ECONOMETRIC ISSUES

Another unresolved issue in this literature relates to timing and causality. In this section, I outline a methodology that helps us examine issues of timing and endogeneity bias in greater detail. I also dis-

**Table 3.6 Distributions of Year of Adoption of AMTs
by the 79 Survey Firms**

Year of adoption	Number of technologies	Percent	Cumulative number	Cumulative percent
1970	2	0.9	2	0.9
1972	1	0.4	3	1.3
1973	1	0.4	4	1.7
1974	1	0.4	5	2.1
1975	4	1.7	9	3.8
1977	1	0.4	10	4.2
1979	3	1.3	13	5.5
1980	8	3.5	21	9.0
1981	5	2.2	26	11.2
1982	3	1.3	29	12.5
1983	11	4.8	40	17.3
1984	22	9.6	62	26.9
1985	27	11.8	89	38.7
1986	21	9.2	110	47.9
1987	24	10.5	134	58.4
1988	45	19.8	179	78.2
1989	24	10.5	203	88.7
1990	25	10.9	228	99.6
1991	1	0.4	229	100.0

cuss some econometric issues that naturally arise in estimating models of skill-biased technological change.

A standard model that is used to test for skill-biased technological change is based on the estimation of a reduced-form version of a cost function in which some proxy for technology is included as an argument. This enables the researcher to test for the non-neutrality of technical change by examining the sign (and significance) of the coefficient on the technology variable. For example, Berman, Bound, and Griliches (1994) tested for capital-skill complementarity based on the estimation of a restricted labor cost function:

Eq. 1 $LC = \mathrm{f}(w_i, K, Q, t)$

where

$LC =$ labor cost
$w_i =$ wage of the ith type of worker
$K =$ the stock of capital
$Q =$ output
$t =$ time

and f is assumed to have a translog form. Invoking cost minimization of the variable inputs, Shephard's lemma (which is $s_i = \partial LC / \partial W_i$), constant returns to scale, and homogeneity of degree 1 in prices, and then taking first differences yields

Eq. 2 $\mathrm{d}s_i = \beta_0 + \beta_1 \mathrm{d}\ln(W_i / W_j) + \beta_2 \mathrm{d}\ln(K/Q) + u,$

where

$s_i =$ the share of labor type i in total employment or labor cost
$K/Q =$ the capital intensity
$u =$ a classical disturbance term[6]

If $\beta_2 > 0$, we have "capital-skill" complementarity, a term coined by Griliches (1969) in his seminal article.

To examine the relationship between labor composition and technical capital, Berman, Bound, and Griliches (1994) included indicators of the intensity of technological investment in Eq. 2, such as the ratio of research capital to output (R/Q). They estimate the following equation:[7]

Eq. 3 $\mathrm{d}s_i = \beta_0 + \beta_1 \mathrm{d}\ln(W_i / W_j) + \beta_2 \mathrm{d}\ln(R/Q) + u.$

We wish to examine whether shifts in labor composition are related to two additional proxies for technological change in the equation

Eq. 4 $\mathrm{d}s_i = \beta_0 + \beta_1 \mathrm{d}\ln(W_i / W_j) + \beta_2 \mathrm{d}\ln(R/Q) + \beta_3 \mathrm{AGETECH}$
$\qquad\qquad + \beta_4 \mathrm{NUMTECHS} + u$

where AGETECH is a measure of the average age (years since being implemented) of a firm's stock of advanced manufacturing technologies and NUMTECHS is the number of technologies a firm implements.

There are several econometric concerns that arise in estimating Eq. 4. One is whether relative wages are exogenous, even at the firm level.[8] Another concern is simultaneity, because the age of technology could also be determined by changes in relative wages and the rate of investment in technology. There may also be substantial multicollinearity among the technology indicators, since they are basically measuring the same phenomenon. Finally, there is also concern regarding measurement error in the reporting, because firms were asked to report the wage data over a four-year period and may have provided us with rather crude estimates for previous years.[9]

In sum, although this may be the appropriate version to estimate theoretically, there is a strong likelihood that several of the basic assumptions of the linear regression model are violated. Thus, I decline to estimate Eq. 4 in favor of a simpler econometric model, which is described in the next section.

PROPOSED ECONOMETRIC SPECIFICATION

A less stylized examination of wage and compositional effects can be based on estimating regressions of the form

Eq. 5 $\ln s_{ijk,t+n} = \beta_1 \, \text{AMT}_t + \beta_2 \ln s_{ijk,t-1} + \gamma_k + u_{ij,t+n}$

where

s = the employment and labor cost shares for the ith class of worker

$ijk,t+n$ = worker class i in firm j in industry k in year $t + n$ ($n = 1, 2, 3$)

AMT_t = 1 if the firm adopted an advanced manufacturing technology between $t - 1$ and t, and otherwise equals zero

γ_k = a "fixed effect"

u = a classical disturbance term

Six classes of workers are observed: managerial and supervisory (MS), technical and professional (TP); clerical and administrative (CL); direct (production) labor and supporting personnel (DPL); other production labor (OTH), mainly service workers; and R&D scientists and engineers (RD). Note that simply comparing the growth rates of employment for adopters versus non-adopters is equivalent to setting $\beta_2 = 1$ and $\gamma_k = \gamma, \forall j$. This specification allows us to test the following hypothesis, which is perhaps the central theme of the monograph.

Hypothesis 1: AMT adoption is associated with corporate downsizing and a shift in labor composition in favor of highly educated workers.

In this version of the model, the variable AMT has the possibility of taking on several values depending on how we decide to disaggregate technologies (an additional subscript can be added). Recall that our data are unique in the level of detail available on AMT and on the labor categories. We have direct, explicit information on 12 types of AMTs. In Chapter 4, I will argue that 10 of these 12 AMTs can be grouped into two distinct classes of technologies, linked and integrated, and we will examine employment, compositional, and empowerment effects separately for each class of technology. Also, while most studies have focused on two measures of labor inputs, production and non-production workers, we can examine the labor force implications of technological change for the six classes of workers outlined above.

Note also that in Eq. 5, including the initial values on the right-hand side is equivalent to relaxing the restriction that $\beta_2 = 1$. Controlling for initial size, average wage, or composition could be important if these variables are jointly determined or correlated with adoption or success after implementation.[10] For instance, firms that tend to have a large share of technical and professional workers may be more adept at identifying new technologies. Similarly, high relative wages for certain groups of workers may induce firms to adopt labor-saving innovations.

Industry controls are also important, because some of the variation in s could be due to industry factors. For example, during the sample period, many manufacturing firms were downsizing, due (in part) to a decline in demand and greater foreign competition. Furthermore, the end of the Cold War resulted in a dramatic downturn in the demand for weapons systems and other equipment to counteract the Soviet threat. When lucrative defense contracts expired without renewal, defense personnel, including engineers, were fired. This, in turn, stimulated a downturn in demand for subcontractors of defense firms.

In the specification of Eq. 5, β_1 is interpreted as the percentage difference in the mean value of the growth rate of the share of employment (for each type of worker) between firms that adopted an AMT between year $t-1$ and t and those that did not. This proposed specification also enables us to examine the timing of labor force adjustments.

I will estimate the factor share models in two ways. The first treats the adoption of a new technology as an exogenous event and simply examines the changes in employment shares (and wages) that result, based on a simple regression model. A second method, based on a two-stage estimation procedure, controls for the endogeneity of these events. In the first stage, I model the firm's decision to adopt the technology, where the probability of adoption is assumed to be a function of a set of industry and firm characteristics, including the firm's experience with related technologies. Having controlled, to some extent, for the endogeneity of technological events, I reinvestigate the employment effects in the second stage. I hypothesize that it is important to include these controls because decisions to adopt technologies and change the composition of the workforce may not be truly independent. In other words, a firm may decide to innovate precisely because the new technology allows them to lower their labor costs and/or adjust the composition of their workforce.

The salient point is that the use of the variable AMT in Eq. 5 raises concerns about endogeneity. There are several econometric techniques that can be used to adjust for this problem. As a first attempt, we propose two-stage probit estimation, as outlined in Lee, Maddala, and Trost (1979) and Maddala (1983). In the probit equations, AMT_i is regressed on a vector of firm and industry characteristics (Z_i), where $AMT_i = 1$ if the firm adopts technology i, and is zero otherwise.

Following Maddala (1983), the model has the following form:

$$\text{Prob}(\text{AMT}_i = 1) = \text{Prob}(u_i \geq -\gamma'Z_i) = 1 - F(-\gamma'Z_i)$$

and

$$\text{Prob}(\text{AMT}_i = 0) = \text{Prob}(u_i < -\gamma'Z_i) = F(-\gamma'Z_i),$$

where the error term of the regression equation, u_i, is assumed to have zero mean and constant variance, σ^2; F is the cumulative density function of the standard normal distribution; and γ' is a parameter vector. We propose to estimate probit equations for groups of related and individual advanced manufacturing technologies.

Next, we consider the appropriate set of firm characteristics, Z. In recent empirical studies of technology adoption (Hannan and McDowell 1984; Dunne 1994; Levin, Levin, and Meisel 1987), Z includes the age, size, and R&D intensity of the firm, as well as an industry dummy. Age is designed to capture possible "vintage" effects, i.e., the possibility that firms having a more recent capital stock would be more likely to adopt new technologies because of "innovational complementarities" (a term coined by Bresnahan and Trajtenberg [1995]) or because of lower adjustment costs.[11] Size is included because large firms may have better access to capital or because projected returns exceed threshold levels for adoption only in the case of large-scale projects.[12] This is consistent with one aspect of the so-called Schumpeterian hypothesis, which postulates that large firms conduct more R&D than small firms. Theoretical models of technology adoption and diffusion (Cohen and Levinthal 1989, Jensen 1988) emphasize the role of R&D investment in helping companies overcome informational and technical uncertainties regarding new technologies. R&D-intensive firms may also have lower costs for training and spend less time and effort on integration activities and other aspects of adjustment to new technologies.

I use an additional covariate, NUMTECHS, that serves as a proxy for a company's experience with related technologies, which I view as an indicator of organizational learning. When a firm implements a new technology, a period of learning and adjustment to the new production process follows. I hypothesize that perfecting one technology reduces the uncertainty and adjustment costs associated with subsequent technological investments. This view predicts higher expected returns, on average, for subsequent adoptions. It also predicts that the probability of adoption is a function of previous experience with related technologies. After deriving parameter estimates of the probit equations, I reestimate Eq. 5 in the second stage.

This two-stage, instrumental variables approach to "exogenize" the technology variable is not without controversy. In this framework, the instruments (firm characteristics in my model) are assumed to be exogenous and thus uncorrelated with the error term in Eq. 5 in the second stage. This is a potential problem, because Bound, Jaeger, and Baker (1995) showed that when the instruments explain only a small fraction of the variation in the endogenous explanatory variable (in this case, technology) in the first stage, the second stage estimates will be subject to large inconsistencies.[13] In Chapter 5, which discusses the empirical results, I will present evidence that supports the use of this method. In sum, there are two hypotheses relating to the determinants of technology adoption that I wish to test:

Hypothesis 2: The probability of technology adoption is positively associated with firm size, R&D intensity, and experience with related technologies.

Hypothesis 3: The probability of technology adoption is negatively associated with the age of the firm.

Note that the comprehensive nature of our data allow us to also perform a dynamic analysis of technology adoption; that is, we can estimate a hazard function, because we observe the firm's complete history of technology adoptions. Several studies of technological diffusion have been based on estimation of a hazard function. In this context, a hazard function is defined as the conditional probability that a firm will adopt a new technology during a year, given that it has not done so by the beginning of that year. Romeo (1975), Hannan and McDowell (1984), and Levin, Levin, and Meisel (1987) estimated dynamic models of the adoption of numerical controllers, ATMs, and optical scanners, respectively.

The estimation of this class of models typically begins with defining the survivor function:

$$\textbf{Eq. 6} \quad S(t) = \exp\left(-\int_0^T h(t)dt\right),$$

where $h(t)$ is the conditional probability that the firm will adopt an advanced manufacturing technology at time t. This hazard function is defined as

$$\textbf{Eq. 7} \quad h(t) = \lim_{\Delta t \to 0} \frac{[P(\text{event occurs in } (t + \Delta t)/(T > t)]}{\Delta t}$$

$$= d[\ln S(t)]/dt$$

$$= f(t)/S(t)$$

where $f(t)$ is the probability density function.

As in Levin, Levin, and Meisel (1987), we specify a proportional hazards model, in which the hazard function is expressed as

$$\textbf{Eq. 8} \quad h(t,Z,\beta) = h_0(t)\exp[\beta'_k Z_k]$$

where h_0 is an unspecified baseline probability and Z is the set of covariates that was defined previously in this chapter. The proportional hazards model offers an important advantage by generating parameter estimates (β) that do not depend on the functional form for $h_0(t)$. Thus, this framework obviates the need to define a specific functional form, such as the exponential or Weibull distribution. One disadvantage of this approach is that we cannot determine the time dependence of the hazard function.

In Chapter 5, I report estimates of the static (probit) and dynamic models.

Notes

1. It is important to note that firms are notoriously reluctant to provide what they usually view to be proprietary information on technology usage and labor composition. Also, it was sometimes difficult to identify the appropriate company official to complete the survey, since filling it out requires extensive knowledge of the technology, manufacturing, and human resource areas of the firm. For smaller firms, this was generally not difficult.

2. Surveys which focused on large companies, such as a random sample of human resource management practices of New York Stock Exchange firms (New York Stock Exchange 1982) or a survey of total quality management (TQM) practices of Fortune 5000 companies (Lawler III, Mohrman, and Ledford 1992), achieved higher response rates, 26.5% in the former and 32% in the latter. A recent establishment-level, national survey of training practices conducted by the U.S. Bureau of Labor Statistics achieved a 72% response rate (Black and Lynch 1996).

3. Source: U.S. Census of Manufactures.

4. In the preface to the survey report, U.S. Census Bureau officials justify conducting the survey by describing how these technologies have dramatically changed production methods. The fact that the Census Bureau believed that such a survey was warranted, even during a period of severely constrained budgets for data collection, is telling.

5. Note that in the survey (see Exhibit 1), we ask whether the technology is "implemented and in use."

6. Berndt, Morrison, and Rosenblum (1992) estimated a variant of this model with other compositional measures as dependent variables.

7. Bartel and Lichtenberg (1987, 1990) estimated the following variant of this model:

$$ds_i = \beta_0 + \beta_1 \, d\ln(W_i/W_j) + \beta_2 \, d\ln(K/Q) + \beta_3 \, \text{AGETECH} + u,$$

where AGETECH is a measure the age of the industry's technology, proxied by estimates of the age of the capital stock. The authors found an inverse relationship between AGETECH and the average wage, labor cost, and employment share of highly educated labor. This evidence is interpreted as being consistent with learning, since the demand for highly educated labor is hypothesized to be highest in industries with relatively young technologies. Also, greater effort, and thus higher wages, are required in the initial stages of implementation.

8. Another problem is that there may not be much variation in wages, particularly for firms in the same region.

9. Strictly speaking, our sample is not a panel dataset, because the survey was conducted at a single point in time.

10. This could be true if larger firms are more likely to adopt technologies than smaller companies or if increases in revenue result from successful implementation.

11. An effect that may counteract this is "survivor bias," as discussed in Dunne, Roberts, and Samuelson (1989); that is, I only observe the older firms that have survived.
12. Theoretical studies by Pakes and Ericson (1989) and Jovanovic (1982) suggest a positive correlation between productivity and firm size, which might be due to a more rapid rate of technological innovation among large companies.
13. I am indebted to Eli Berman for pointing this out.

4 Characteristics of Advanced Manufacturing Technologies

The purpose of this chapter is twofold: first, to provide the reader with background information on advanced manufacturing technologies (AMTs), and second, to show that it is important to distinguish between two classes of technologies, linked and integrated. Finally, I discuss how changes in the organizational environment may arise from implementation of AMTs, resulting in greater employee empowerment.

AMTs typically involve substantial investment in new hardware and software, which necessitates investment in retraining. Some scholars have interpreted this as a form of "skill upgrading." Also, AMTs are designed to achieve integration across functional activities (marketing, manufacturing, R&D, accounting/finance, logistics, purchasing, and product design).

A greater emphasis on integration could lead to dramatic changes in methods of production, which would have important implications for productivity measurement and labor market conditions; that is, AMT implementation often results in a new production process (often interpreted as a shift in the cost function) and could lead to a shift in the demand for labor. Many firms report a greater emphasis on teamwork and cooperation (between union and management) after implementation.

AMTs have had a strong impact on manufacturing performance. Studies by Chen and Adam Jr. (1991) and McGuckin, Streitweiser, and Doms (1998) reported that AMT adoption is associated with higher productivity. AMTs have also changed our traditional concept of a trade-off between cost and quality. Evidence suggests that AMTs provide a means by which manufacturing processes can achieve higher quality without sacrificing delivery and flexibility performance and with only slightly higher costs. MacDuffie, Sethuraman, and Fisher (1996) compared "lean" production plants (those which use one or several AMTs) with traditional mass production plants in the automobile industry. They found that plants that have implemented these new technologies have higher levels of product variety than traditional plants. Furthermore, the lean plants can produce different types of products without significantly reducing labor productivity.

Other studies show that AMTs have resulted in a reduction in lead times, improvements in product quality, and a reduction in the probability of plant failure (Chen and Adam Jr. 1991; Doms, Dunne, and Roberts 1994). Dunne (1994) and Dunne and Schmitz (1995) reported that AMT usage enhances the quality of labor, as reflected in a higher average hourly wage and an increase in worker skill levels.

While these studies have been useful, they did not examine the effect of different classes of technologies on labor market outcomes. I believe that, in this context, it is important to distinguish between linked and integrated AMTs. Specifically, I hypothesize that companies may phase their implementation of AMT by first linking design and manufacturing efforts and then integrating the manufacturing enterprise. These stages and their potential relationship to downsizing, compositional, and other human resource management strategies (such as employee empowerment) are described in the following sections.

LINKED VS. INTEGRATED TECHNOLOGIES

Linked AMTs

The first phase of AMT adoption involves informational linking of the design and manufacturing functions and the establishment of quality-control and production-control practices. By feeding designs from engineering directly to the shop floor, manufacturers can speed design, decrease product development and coordination costs, increase flexibility in response to customer order changes, and maintain consistently high levels of quality.

Computer-aided design (CAD) and computer-aided manufacturing (CAM), jointly referred to as CAD/CAM, provide a mechanism for sharing product design and process-control information between the design and manufacturing groups. The operations management literature (see Chase and Aquilano 1995) suggests that while CAD and

CAM can be used as stand-alone technologies, the real benefits occur as a result of <u>linking</u> the two technologies.

Computer-aided design involves the design of products using graphical computer software containing databases of standard parts. The typical design activities that occur through a CAD system are preliminary design of a new product, drafting, modeling, and simulation. There are numerous examples of the use of CAD by firms in a wide variety of industries. Standard-part databases allow General Motors and Ford to use CAD tools to design, source, and manufacture components in different countries (Davenport 1993). Proctor and Gamble used this technology to design a pump dispenser for Crest toothpaste (Heizer and Render 1999). B.F. Goodrich engineers model wheel and brake assemblies using a CAD system. Goodrich's experience with CAD illustrates that implementation of this technology can also be used to reduce costs. The firm reported that when conditions of stress and heat are simulated on the computer, engineers can identify and remedy errors at the design stage on a computer screen, instead of fixing them a much higher cost after a new product line has been launched (Heizer and Render 1999).

As discussed in Chase and Aquilano (1995), CAD can directly connect to computer-aided engineering (to analyze engineering characteristics) and computer-aided process planning (to generate the parts programs fed to computer-controlled machine tools). Perhaps the most salient benefit of CAD is the ability to reduce lead times in product development. Many of the companies that I visited, such as Symbol Technologies, a leading manufacturer of bar-code scanners, have been able to introduce a wide array of new products in a short period of time because of the use of CAD technology.

Computer-aided manufacturing involves the use of computers in the planning, management, control, and operations of a manufacturing facility. Computer applications relating to CAM include inventory control, scheduling, machine monitoring, and processing of all information relevant to the manufacturing process. The primary use is for transferring, interpreting, and monitoring manufacturing data. The key advantage of CAM is the ability to pre-program production processes and to rapidly change from one part program to another. Early adopters of CAM used centralized computers to store information on parts, whereas more recent configurations of CAM allow for locally controlled computers (computer numerical control [CNC]).

As described in Schroeder (1993) and Cohen and Apte (1997), Boeing's Commercial Airline Group is an avid user of CAD/CAM technology. Boeing has found that integrating design and manufacturing is essential to keeping up with stiff foreign competition from its rival Airbus, because it can significantly reduce development costs and raise manufacturing quality. Specifically, they report shorter product development lead times, greater reliability, improved maintainability, and cost-effectiveness. The Boeing 777 was the first Boeing plane produced that was completely designed on computer and the single largest example of the use of CAD/CAM in manufacturing. CAM integrates and provides useful information to all functional areas of the business: manufacturing, logistics, quality control, marketing, human resources, R&D, financial, and legal functions. It often serves as a critical focal point for the receipt of information from each of these areas and for the transmission of appropriate instructions to each area.

A CNC tool usually has a small computer dedicated to it, so that programs can be stored and developed locally. It uses a computer, instead of an operator, to control a manufacturing process. Computer control facilitates the addition of feedback sensors to machines to monitor part wear and alignment (Adler 1988; Ayres 1988). A substantial percentage of the world's machinery that is engaged in drilling, boring, and milling is now designed for computer numerical control. Many machine tools are based on CNC technology. CNC machines were a precursor to robotics.

The increased power of computers and the application of computer networks created the necessary capabilities to link CAD and CAM. CAD/CAM, with its embedded computer-aided process planning, permits the flow of designs to CNC machines on the shop floor. This link facilitates rapid prototyping, flexibility in responding to customers' requests, and reduced product development and delivery times. Additionally, the database of part specifications, combined with computer control, can be used to compare acceptable dimensions with actual process performance to monitor both product and process quality through the use of statistical process control (Juran 1989). For example, feedback sensors on CNC machines automate data collection, while linked computers create real-time control charts for monitoring processes.

Statistical process control (SPC) enables operators to quickly recognize when their processes are out of adjustment. This allows the manager to identify sources of error and to focus problem-solving efforts accordingly.

Many manufacturing and service firms use SPC. The General Motors Delta Engine plant uses SPC to ensure product quality and reliability. Process control charts are also used by McDonalds to guarantee consistency in taste (Markland, Vickrey, and Davis 1998). Finally, BIC employs SPC to test ballpoint pens for acceptable performance. Many of the companies in our sample reported a significant improvement in quality control after the implementation of SPC.

Despite high fixed start-up costs, linked AMTs should yield lower variable costs than conventional technologies. Costly product development efforts are streamlined and the ability to manufacture high-quality products is enhanced through CAD/CAM. Kelley (1994) found that the flexibility of CAD/CAM reduces the costs of both large and small batch manufacture. Furthermore, CAD/CAM facilitates the application of SPC to maintain quality control of production systems by facilitating self-monitoring, self-regulation, and self-correction (De Pietro and Schremser 1987).

Just-in-time (JIT) production, another AMT, is a production-control system guided by a waste-reduction philosophy, in which work-in-process inventories are continually reduced (Kusiak 1985; Schlie and Goldhar 1989; Schonberger 1986). JIT technologies are often especially effective for repetitive operations. Quality control is closely associated with JIT because defective parts in production processes disturb the continuous flow of product. Thus, defective or low-quality inputs become more evident to workers. The implementation of a JIT system means there are fewer materials in the work flow, and thus the amount of work space needed is reduced. In sum, JIT is specifically designed to reduce manufacturing setup times, variability, and inefficiency in materials/inventory management.[1]

Toyota is generally regarded as the pioneer of JIT techniques. Not surprisingly, when General Motors teamed up with Toyota to build the New United Motor Manufacturing, Inc. (NUMMI) plant in Fremont, California (best known for producing Geo Prizms), JIT methods were implemented. As described in Markland, Vickrey, and Davis (1998), plant managers instituted quick delivery arrangements with suppliers in the Midwest. On a daily basis, the plant receives exactly enough parts to produce 900 vehicles (one day's production quota). This one-day supply is consistent with the just-in-time philosophy. When General Motors announced that it was building its new Saturn Division in Spring Hill, Tennessee, its suppliers announced that they were establishing operations nearby. In this case, proximity was critical because it was anticipated that Saturn would require frequent, small-quantity deliveries. Other firms that have adopted JIT methods include Dell Computers, L. L. Bean, Ford, and Caterpillar.

In sum, I hypothesize that linked AMTs involve a combination of computer-aided design (CAD), computer-aided manufacturing (CAM), computer numerical control (CNC), statistical process control (SPC), and just-in-time (JIT) technologies. The recent literature has characterized these technologies as collectively being the forerunner of more modern, integrated technologies. For example, Ettlie and Reza (1992, p. 806) referred to CAD and CNC as "first-generation technology deployment strategies" that newer, "integrated-flexible" technologies could render obsolete.   In the next section, I describe these integrated AMTs.

Integrated AMTs

Despite the advantages of linked AMTs, they are only the first phase in automating the manufacturing process. Integrated AMTs provide a more comprehensive form of automation and greater benefits. For example, flexible manufacturing systems (FMSs) integrate CAD, CNC work centers, and automated material handling and control systems. FMSs provide flexibility because the operations performed at each work station, as well as the routing of parts among work stations, can be directed through software controls.

Matsushita Electric Industrial Company (best known to Americans as the producers of Panasonic equipment) has an FMS facility in Japan that produces VCRs. As reported by Chase and Aquilano (1995), this facility features a highly automated robot assembly line consisting of about 100 work stations. With the exception of a few troubleshooters and engineers, Matsushita can operate this facility using very few workers. Furthermore, their FMS enables them to produce any combination of over 200 models of VCRs.

An FMS can range in complexity from a numerically controlled machine with an automated pallet changer and parts buffer, to complicated networks of manufacturing cells connected through automated material handling systems, robots, and computer networks, as in the case of Matsushita (Adler 1988; Skinner 1974). Chen and Adam Jr. (1991) noted that a key purpose of FMS is to efficiently manufacture multiple parts at low to medium volumes. As described in Schroeder (1993), Lockheed-Martin and Grumman-Northrop both use FMS to produce many of the small parts (brackets, clips, spacers, fillets, and gussets) that constitute most of the small sheet metal parts in military aircraft, such as F/A-18 Hornet strike fighter. They report substantial reductions in manufacturing lead time and work-in-process inventory, as well as improved product quality.

Automated guided vehicles (AGVs) and automated storage and retrieval (AS/R) systems provide an integrated means for automated movement of parts and raw materials from storage areas to the cells of an FMS and from cell to cell. AS/R systems are high-density storage systems in which computers control automatic loaders that pick and place items to and from storage (Zygmont 1987). These systems are a key element in a computer-aided materials-handling system, providing an automated means of tracking work-in-process inventory. AS/R systems tend to be quite expensive, but they are available in a wide variety of sizes and levels of specification. Heizer and Render (1999) report that Wal-Mart, Benetton, and Tupperware have large computer-controlled warehouse and distribution facilities that are based on AS/R technology.

Automated guided vehicles can be used in conjunction with AS/R systems to automate and integrate material handling throughout the plant. AGVs travel under radio control or by following a small strip embedded in the plant floor. Automatic loaders retrieve parts, components, and assemblies from storage racks, and AGVs deliver the items to the appropriate work center on the factory floor. AGVs can also transport items between work centers.

An AGV is a special type of robot, and robotics (ROB) is important in integrated AMT facilities. The Robot Institute of America (1986, p. 1) defines a robot as a "reprogrammable, multifunctional manipulator designed to move material, parts, tools or specialized devices through variable motions for performance of a variety of tasks." Robots are especially useful when environmental hazards create dangerous working conditions; for example, robots are used extensively by automobile manufacturers in painting and welding operations. Monotonous tasks, such as inserting a screw or metal-cutting, are also likely candidates for robots to perform. Robots play an important role in the automated factories of IBM (Proprinters), General Motors (Saturn), and American Standard (painting). In a recent study of Fortune 500 companies, firms cited increased production, improved quality, and increased safety as primary reasons for the implementation of robotics (Schonberger 1982).

A study by Chase and Aquilano (1995) indicates that the development of group technology (GT) was critical to the success of flexible manufacturing systems, because it is impractical to design an FMS cell to produce an unlimited range of part types. Group technology is based on the concept of identifying and synthesizing related attributes to achieve efficiencies by "grouping" similar problems.

In the context of FMS, group technology is a parts coding scheme that develops "part families" based on dimensions, shape, material, tolerances, and other specifications. It is best suited for small-batch, high-variety manufacturing. An FMS layout based on part families results in dramatically less travel and idle time than is the case in functional (departmental) layouts (Adler 1988; Ettlie and Reza 1992; Hollen and Rogol 1985; Robot Institute of America 1986). It is clear that group technologies are integrative.

In sum, I hypothesize that integrated AMTs are composed of a combination of flexible manufacturing systems (FMSs), automated storage and retrieval (AS/R) systems, automated guided vehicles (AGVs), robotics (ROB), and group technology (GT).

In this chapter, I have argued that it is important to distinguish between two classes of advanced manufacturing technologies (AMTs), linked and integrated. The first provides an informational "link," through computers, of the design and manufacturing activities, which leads to an enhancement of quality control and control of the entire manufacturing process. The second enables the firm to "integrate" a wide range of activities in the manufacturing facility in order to streamline efficiency by lessening the need for labor, capital, and materials.

The material presented in this chapter suggest the following hypotheses, which underscore the importance of conducting a disaggregrated analysis of skill-bias technological change:

Hypothesis 4: We can distinguish between two classes of AMTs, linked and integrated.

Hypothesis 5: The association between integrated AMTs and downsizing will be stronger than the association between linked AMTs and downsizing.

Hypothesis 6: The "skill-bias" of technological change will be stronger for integrated AMTs than for linked AMTs.

EMPLOYEE EMPOWERMENT ASPECTS OF AMT

Empowerment is the process of giving lower-level employees decision-making power. Workers often respond well to being given greater autonomy and responsibility. Greater empowerment may lead to an increase in productivity and product quality, in part, because empowerment typically enhances employee motivation. Several studies have documented such performance effects (Conger and Kanungo 1988; Thomas and Velthouse 1990). In the literature on technology and innovation management (Adler 1988), employee empowerment is considered to be a critical component of successful AMT implementation.

Appelbaum and Batt (1994) provided a comprehensive review of empowerment initiatives (what they term "employee involvement" programs) at major U.S. corporations. They found that Japanese "lean production" methods, which often involve the use of linked AMTs such as statistical process control (SPC) and just-in-time (JIT) inventory systems, are designed to simplify the production process. Employee involvement in quality improvement, through quality circles, is crucial to the success of these programs. Quality circles consist of groups of employees from specific work areas who volunteer to meet on a regular basis and provide suggestions to upper management for improvements in quality and/or productivity. The authors note that, in practice, quality circles may not really empower workers all that much, because these groups often do not have any explicit power beyond an advisory role and there are usually no rewards for participation. On the other hand, they also reported that lean production methods are associated with formal "power sharing" between workers and managers through "joint consultation committees," which do have some decision-making power.

Thomas and Velthouse (1990) note that there is no broad-based consensus on how to construct firm-level measures of empowerment. Still, most existing measures (see Daft 1998) have been based on a composite of several "intervention methods," which serve as indicators of a firm's efforts to empower workers. These intervention methods could include such factors as training, changes in job responsibilities, and the creation of new jobs and career opportunities for existing workers. They are considered legitimate proxies for empowerment because they potentially enhance worker motivation and provide employees with the requisite skills to make more complex decisions. Managers support these efforts because they believe that empowered workers will have the self-confidence, freedom, and motivation to continuously solve problems.

Many firms, such as Chrysler and Hampton Inns, use training as a means to empower workers and achieve continuous improvement (Daft 1998). For example, plant managers at Chrysler's manufacturing facilities regularly present quality awareness workshops for production workers, which has encouraged employees to initiate many useful suggestions for quality improvements. Employees of Hampton Inns are provided with comprehensive, ongoing training that reinforces the rationale behind the firm's commitment to its number one strategic goal, 100% customer satisfaction. Workers have been empowered by senior management to do whatever is required to achieve this goal. Hampton also uses career advancement opportunities and pecuniary incentives to motivate and reward workers who upgrade existing standards of customer service.

In constructing my measure of empowerment, I follow the advice of Conger and Kanungo (1988), who contend that such measures should be based on the aforementioned intervention methods and the key empowerment outcome of increased employee "control," which is defined as greater decision-making power. This too has a motivational component, in that control is associated with greater responsibility and accountability.

How might the adoption of AMTs be associated with employee empowerment? A number of studies have pointed to an increase in training and a stronger commitment to learning in the aftermath of AMT adoption (Adler 1988). For example, proponents of statistical process control (SPC) have emphasized the importance of training to accompany its implementation (Klein 1991). One company that has demonstrated a strong commitment to learning in an AMT environment, through employee empowerment, is Chaparral Steel, listed by Fortune magazine as one of the 10 best-managed factories in the United States. A case study of the firm by Leonard-Barton (1992) cited its factory as a "learning laboratory," where each worker has been empowered to engage in R&D. One notable feature of Chaparral's empowerment policy is its compulsory annual education leave program that requires employees to visit customers, "best practice" companies, and universities to learn new processes and technologies. This openness to knowledge from outside the firm stands in sharp contrast to the conventional "not invented here" syndrome, or resistance to new ideas from external sources, that pervades many companies.

Several empirical studies also report a positive effect on job enrichment, increased job opportunities and roles, and increased employee control. For example, AMT appears to be associated with enlarged job responsibilities and the creation of new roles for employees (Adler 1988). In a theoretical paper, Klein (1991) argued that AMT might increase employee control, although such control may be limited to task design as opposed to task execution. In sum, I wish to test the following hypothesis:

Hypothesis 7: AMT adoption enhances employee empowerment.

Perhaps the most comprehensive study of the human resource management implications of AMT was conducted by Snell and Dean (1992). The authors contended that workers must be trained and empowered to handle the increased complexities, judgment, and problem solving in an AMT environment. They found a strong, positive connection between the adoption of AMTs and comprehensive training and human resource development strategies. However, their AMT measure, derived from a factor analysis, was oriented towards linked technologies such as SPC and JIT. The authors did not measure the human resource effects of integrated technologies such as robotics and flexible manufacturing systems.

Although all these studies suggest a positive association between AMT and employee empowerment, the relationship may depend on the type of AMT that is implemented, because the studies focused largely on linked technologies. In the remainder of this chapter, we consider the human resource and employee empowerment implications of integrated technologies.

To illustrate the differences in employee involvement for linked vs. integrated AMTs, consider a JIT system. JIT is based on a philosophy of continuous incremental improvement in the manufacturing process. As noted in Appelbaum and Batt (1994), the success of JIT depends to a large extent on increasing employee involvement. This worker input enables managers to reduce time and inventory "buffers" (inefficiencies) between work stations and between a vendor and the plant. A better awareness by workers of the needs and problems of their co-workers and customers, along with a more cooperative work environment, leads to a significant improvement in quality and a reduction of waste.

Contrast this scenario to the one that arises in the aftermath of the implementation of an integrated AMT, such as a flexible manufacturing system. An FMS can result in the use of certain types of automation, such as numerically controlled machines or robots, or in factories that are entirely automated, such as General Electric's Aircraft Engine Group in Lynn, Massachusetts. Many firms implement FMSs with the addition of "expert systems," which are computer programs that simulate the behavior of human experts. These knowledge-based systems allow for substantial improvements in productivity because they reduce the need for human input. They are often used in conjunction with an initiative to enhance factory automation.

Given the characteristics of the respective classes of technologies, I hypothesize that the extent of employee empowerment will be lower with integrated AMTs than with linked AMTs. The automated nature of integrated AMT will not enhance the training, skill enhancement, job responsibilities/opportunities, or control on the part of existing workers. Instead, control will be largely in the hands of programmable machinery and a select group of skilled, technical personnel. Work by De Pietro and Schremser (1987) corroborates these assertions, finding that the introduction of robotics did not have a significant effect on job influence or control. In sum, I wish to test the following hypothesis:

Hypothesis 8: The extent of employee empowerment will be greater for linked AMTs than for integrated AMTs.

The Long Island survey contains several important indicators of the degree to which employees are "empowered" by technological change. As shown in Exhibit 1, respondents were asked detailed questions regarding the methods of AMT implementation. Following Conger and Kanungo (1988), I used three intervention methods and one outcome measure to construct a measure of employee empowerment. Listed in the survey under methods of implementation, the three intervention methods are

- training of existing personnel,
- changing employees' job responsibilities, and
- creating new jobs and career opportunities for employees,

to which I add the key empowerment outcome of

- increased employee control (which is listed in the survey under results of implementation).

Thus, I can determine the extent to which AMT adoption leads to an increase in each of these four factors.

A DISAGGREGATED ANALYSIS OF SKILL-BIASED TECHNOLOGICAL CHANGE

Existing studies of skill-biased technological change typically treat technology as a homogeneous activity. This is unfortunate, since the nature of the skill bias could depend on the type of technology that is implemented. There is, in fact, substantial heterogeneity among technologies in terms of their impact on workers. Thus, a disaggregated analysis of skill-biased technological change could provide a more realistic and accurate portrayal of the shifts in labor composition, relative compensation, and employee empowerment that arise when companies adopt new technologies. This evidence could be useful to managers who formulate HRM policies and strategies and to policymakers, to help target subsidies for training programs and retraining displaced workers more effectively.

In this chapter, I have argued that it is appropriate to separately consider the labor market consequences of adopting two <u>classes</u> of technologies, linked and integrated AMTs. I have outlined important differences in their characteristics, skill requirements, and effects on organizations. One of these differences is that the nature of the recompositional shifts in favor of highly educated workers will be greater for integrated AMTs than for linked AMTs. That is simply because integrated AMTs are often implemented with the specific goal of streamlining manufacturing efficiency through reductions in the use of production labor. Recall that integrated AMTs are the most advanced manufacturing technologies and typically require substantial scientific and engineering input. They automate materials handling and control, a function that usually requires large number of low-skilled employees. Furthermore, integrated AMTs are usually adopted by firms that have implemented linked AMTs several years earlier. We hypothesize that this transition from linked to integrated AMT will further reduce the demand for production and clerical workers. Recall also that a primary goal of many linked AMTs is to improve product quality, variety, and reliability, not necessarily a desire to improve productivity. The bottom line is that we expect the skill bias to be more pronounced for integrated AMTs.

We also expect differential effects for employee empowerment. Specifically, we hypothesize that the extent of employee empowerment will be lower for integrated AMTs than for linked AMTs. The automated nature of integrated AMTs will not enhance the requisite training, job responsibilities/opportunities, or control on the part of existing workers. Instead, control will be largely in the hands of programmed machinery and a select group of highly skilled technical personnel. Alternatively, as alluded to in previous sections of this chapter, workers must be trained and empowered to handle the increased complexity, judgement, and problem solving skills required in a linked AMT environment.

It is also important to note that a disaggregated analysis of skill-biased technological change would be consistent with a long-standing literature on assessing the returns to R&D investment. Many of these studies decompose R&D by source of funds (company vs. federally funded), character of use (basic vs. applied research and development), and type (product vs. process innovation). The authors usually report statistically significant differences in the private and social returns for these categories of R&D investment (Link 1987; Lichtenberg and Siegel 1991; Griliches 1998). These studies have important policy and managerial implications because they yielded fresh insights into the economic motives and consequences of investment in R&D. In a similar vein, I hope that an investigation of different classes of technologies will result in a richer understanding of the consequences of technological change on the demand for labor.

Note

1. JIT often replaces a material requirements planning system (MRP) as the primary production-control technology. However, MRP can still be used effectively to manage master production schedules and the ordering of parts and material from vendors.

5 Empirical Results

In Chapter 2, I outlined the theory of skill-biased technological change and reviewed existing empirical studies of this theory. In Chapter 3, I discussed some econometric problems inherent in empirical estimation of models. Researchers often use proxies for technological change (such as R&D investment or expenditures on computers) and have limited information on labor composition. Furthermore, they have great difficulty controlling for the endogeneity of technological change.

The dataset used in this study contains richer information on technological change and workforce composition. It provides us with direct, firm-level measures of technological change, e.g., actual implementations of AMTs and detailed information on labor composition and other relevant firm characteristics. We use these data to examine the antecedents and labor market consequences of technology adoptions.

Before discussing the empirical findings, the following is a summary of the hypotheses presented in previous chapters:

Hypothesis 1: AMT adoption is associated with corporate downsizing and a shift in labor composition in favor of highly educated workers (p. 38).

Hypothesis 2: The probability of technology adoption is positively associated with firm size, R&D intensity, and experience with related technologies (p. 41).

Hypothesis 3: The probability of technology adoption is negatively associated with the age of the firm (p. 41).

Hypothesis 4: We can distinguish between two classes of AMTs, linked and integrated (p. 53).

Hypothesis 5: The association between integrated AMTs and downsizing will be stronger than the association between linked AMTs and downsizing (p. 53).

Hypothesis 6: The "skill-bias" of technological change will be stronger for integrated AMTs than for linked AMTs (p. 53).

Hypothesis 7: AMT adoption enhances employee empowerment (p. 55).

Hypothesis 8: The extent of employee empowerment will be greater for linked AMTs than for integrated AMTs (p. 57).

Hypothesis 1 is perhaps the central theme of the monograph: that technology adoption leads to a substitution of capital (especially in the form of computers) for low-skilled labor and a concomitant shift in demand in favor of more highly skilled workers. Hypotheses 2 and 3, while not directly related to skill-biased technological change, are relevant because of the nature of the econometric specification: in our two-stage model, we estimate the determinants of technology adoption in the first stage. Hypothesis 7 is based on the concept that technology empowers workers by increasing their ability to control work activities and by providing greater autonomy. I find that in the aftermath of technological change, companies respond by increasing expenditures on training, by creating additional career opportunities for workers, and by changing job responsibilities. Each of these changes serves to enhance employee empowerment.

The remaining hypotheses are based on the concept that we can distinguish between two classes of AMTs, linked and integrated. Specifically, downsizing and recomposition of the workforce in favor of highly skilled workers will be greater for integrated AMTs, while the employee empowerment effects will be greater for linked AMTs.

DETERMINANTS OF AMT ADOPTION

The first set of empirical results is the probit estimates of the adoption of selected technologies, based on the model outlined in Chapter 3, "Proposed Econometric Specification" (p. 37). The findings, shown in Table 5.1, are consistent with those reported in Dunne (1994) for the national AMT sample: age and the probability of technology adoption are uncorrelated, and there is a positive relationship between technology adoption and firm size. The results are also consistent with those of previous studies of CNC-related technologies (Romeo 1975; Kelley and Brooks 1991), which indicated that firm size and the probability of technology adoption are positively correlated. The positive and significant coefficient on R&D intensity implies that R&D may help companies acquire knowledge or overcome knowledge-related uncertainties regarding new technologies (McCardle 1985; Cohen and Levinthal 1989; Jensen 1988). Note that the signs and magnitudes of the regression coefficients are relatively similar for different types of technologies. In sum, the evidence appears to confirm Hypothesis 1 and refute Hypothesis 2.

Because the dataset includes a complete historical profile of a firm's investment in AMTs, I can identify the year of AMT implementation. I can also determine whether the firm had made previous AMT investments, which I include in the model as a dummy variable. The positive and significant coefficient on this variable can be interpreted as being consistent with a "learning" process. Furthermore, this appears to be one of the strongest predictors of technology usage. Note that the values of the parameter estimates do not vary substantially across technologies.

In Table 5.2, I present the "dynamic" results, for which the dependent variable is the conditional probability of adoption. The overall pattern of results is quite similar; that is, firm size, R&D intensity, and experience with other AMTs all speed up the process of technological change. However, note that most of the estimates of the magnitudes of these effects are lower in the dynamic framework. This result is consistent with the study of Romeo (1975) that compared regression results from static and dynamic models of the determinants of the adoption of computer numerical controllers.

Table 5.1 Determinants of the Probability of AMT Adoption (probit estimates)[a]

	CAD[b]	CAM	SPC	ROB	JIT	Any linked AMT	Any integrated AMT
Intercept	-1.334***[c]	-1.518***	-1.9789***	-2.4340***	-1.9830***	-1.210***	-1.710***
	(0.4585)[d]	(0.3244)	(0.6456)	(0.7546)	(0.6721)	(0.5416)	(0.5914)
Age	0.0232	0.3241	-0.1245	0.0187	0.0102	0.0619	0.0413
	(0.0124)	(0.24435)	(0.1023)	(0.1293)	(0.0112)	(0.0913)	(0.0400)
Size	0.3421**	0.2410	0.1517	0.1024***	0.2289**	0.1947***	0.1365***
	(0.0835)	(0.1021)	(0.0854)	(0.0442)	(0.1042)	(0.0831)	(0.0548)
R&D intensity	0.5672***	0.7452***	0.3657***	0.8341***	0.7129***	0.6730***	0.7001***
	(0.1976)	(0.2568)	(0.1138)	(0.3569)	(0.2789)	(0.2832)	(0.3102)
Previous adoption	0.4789***	0.6235***	0.3256***	0.7890***	0.7200***	0.5601***	0.6732***
	(0.1864)	(0.2492)	(0.1325)	(0.3378)	(0.2572)	(0.2071)	(0.3004)
ln (Likelihood)	-56.123	-54.236	-52.678	-57.012	-53.012	-55.169	-59.146

[a] Includes industry controls.
[b] CAD = computer-aided design.
CAM = computer-aided manufacturing.
SPC = statistical process control.
ROB = robotics.
JIT = just-in-time production.
[c] *** = statistical significance at the 1% level.
  ** = statistical significance at the 5% level.
[d] Standard errors in parentheses.

Table 5.2 Determinants of the Conditional Probability of AMT Adoption (hazard function)[a]

	CAD[b]	CAM	SPC	ROB	JIT	Any linked AMT	Any integrated AMT
Intercept	−0.986****[c]	−1.101***	−2.234***	−1.569	−1.469***	−1.537***	−1.723***
	(0.3539)[d]	(0.4564)	(0.9846)	(0.8987)	(0.5289)	(0.7459)	(0.4377)
Age	0.0312	0.4232	0.0456	0.0231	0.0324	0.0620	0.0181
	(0.0234)	(0.3467)	(0.0952)	(0.0478)	(0.0197)	(0.0946)	(0.0231)
Size	0.2645***	0.1021	0.1211***	0.0798**	0.1946***	0.1402***	0.2528***
	(0.1012)	(0.0867)	(0.0458)	(0.0349)	(0.0784)	(0.0478)	(0.1182)
R&D intensity	0.4478**	0.3865**	0.2867***	0.7431***	0.6348***	0.6411***	0.5333***
	(0.2021)	(0.1876)	(0.1232)	(0.2879)	(0.3121)	(0.2273)	(0.2082)
Previous adoption	0.3753***	0.5342***	0.5671***	0.4823***	0.6735***	0.5402***	0.6223
	(0.1234)	(0.2212)	(0.1987)	(0.2019)	(0.2749)	(0.2014)	(0.3055)
ln (Likelihood)	−52.321	−53.432	−55.723	−54.136	−56.212	−56.232	−54.245

[a] Includes industry controls.
[b] CAD = computer-aided design.
CAM = computer-aided manufacturing.
SPC = statistical process control.
ROB = robotics.
JIT = just-in-time production.
[c] *** = statistical significance at the 1% level.
** = statistical significance at the 5% level.
[d] Standard errors in parentheses.

EMPLOYMENT AND EMPOWERMENT EFFECTS
OF AMT ADOPTION

The initial set of findings on the employment and compositional effects of AMT adoption are presented in Tables 5.3 and 5.4. Table 5.3 presents mean and median (logarithmic) changes in total employment and employment shares over the sample period for firms that adopted new technologies. These results should be interpreted with caution because they do not include industry controls. (In Chapter 3, I noted that it may be important to include these controls, because downsizing was occurring in some of these sectors due to the downturn in the demand for defense goods.) Still, the findings do suggest a shift towards workers with higher skills and/or education. That is, Table 5.3 shows that technology adopters, on average, experience a decline in total employment and that the distribution of employment shifts in favor of workers with higher levels of education and skill. On average, firms that adopted at least one AMT experienced a 5.8% decline in employment over the sample period. For technology adopters, the fraction of managerial and supervisory, R&D, and technical and professional employees all increased.

These findings are consistent with several recent empirical studies. As reported in Berman, Bound, and Griliches (1994), the manufacturing sector has recently witnessed downsizing and skill-upgrading of the workforce, producing a substantial increase in the level of education and skill of production and nonproduction workers. Brynjolfsson, et al. (1994) found that investments in advanced information technologies were associated with decreases in average firm size. Carlsson, Audretsch, and Acs (1994) and Dunne and Schmitz (1995) reported similar results at the industry and plant level, respectively.

It is important to note that these results are inconsistent with those presented in Van Reenen (1997), who examines a detailed, firm-level panel dataset for publicly traded manufacturing firms in the United Kingdom. After carefully controlling for firm and timing effects, he reports a strong positive correlation between technological change and employment growth. The data on new technologies are derived from the Science Policy Research Unit (SPRU) database, which contains detailed information on successful commercial innovations in Great

Table 5.3 Employment Levels, Changes, and Shares for Adopters vs. Non-Adopters of AMTs[a]

Number of AMTs used	Total employment Mean (median) 1987	1990	Mean (median) change (%)	Mean (median) employment shares (%) MS[b] 1987	1990	TP 1987	1990	R&D 1987	1990	CL 1987	1990	DPL 1987	1990	OTH 1987	1990
0 (N=13)	180 (93)[c]	191 (160)	+12.5 (8.1)	12.1 (8.3)	12.1 (9.7)	6.3 (3.4)	7.0 (3.9)	2.0 (1.7)	3.1 (1.6)	11.7 (11.3)	12.9 (10.7)	70.6 (75.9)	66.8 (67.5)	0.9 (0.0)	1.2 (0.0)
≥1 (N=66)	777 (189)	589 (160)	–5.8 (–7.1)	10.1 (9.5)	11.2 (10.5)	8.9 (9.5)	14.1 (13.1)	6.2 (2.9)	6.7 (3.3)	16.2 (16.0)	13.5 (12.3)	62.8 (66.3)	60.0 (66.7)	2.1 (0.0)	1.3 (0.0)

[a] No industry controls.
[b] MS = managerial and supervisory.
TP = technical and professional.
R&D = research and development.
CL = clerical and administrative.
DPL = direct production labor and supporting personnel.
OTH = other production labor.
[c] Medians are in parentheses.

Table 5.4 Estimates of Differences in Mean Logarithmic Changes between Firms Adopting an AMT in Year t and Those That Did Not (estimates of β_1 from Eq. 5)[a]

	MS[b]	TP	R&D	CL	DPL	OTH
Change in employment shares						
Year $t + 1$	0.062	0.052	0.046	-0.036	-0.062**[d]	-0.045***
	(1.92)[c]	(1.67)	(1.28)	(1.46)	(2.01)	(2.51)
Year $t + 2$	0.102**	0.095**	0.052	-0.045	-0.143***	-0.078***
	(2.04)	(2.03)	(1.35)	(1.62)	(3.32)	(2.82)
Year $t + 3$	0.143***	0.092**	0.084	-0.034	-0.134***	-0.063***
	(2.43)	(1.99)	(1.81)	(1.11)	(3.24)	(2.05)
Change in employment cost shares						
Year $t + 1$	0.059	0.070	0.057	-0.052	-0.072***	-0.051***
	(1.81)	(1.93)	(1.52)	(1.68)	(2.15)	(2.42)
Year $t + 2$	0.127***	0.088**	0.068	-0.061	-0.139***	-0.069***
	(2.12)	(1.96)	(1.60)	(1.71)	(3.12)	(2.73)
Year $t + 3$	0.139***	0.099**	0.096**	-0.049	-0.150***	-0.059**
	(2.21)	(2.03)	(1.95)	(1.32)	(3.19)	(1.97)

[a] Each regression includes an industry dummy.

[b] MS = managerial and supervisory.
TP = technical and professional.
R&D = research and development.
CL = clerical and administrative.
DPL = direct production labor and supporting personnel.
OTH = other production labor.

[c] t-Statistics are in parentheses.

[d] *** = statistical significance at the 1% level.
** = statistical significance at the 5% level.

Britain between 1945 and 1983. Thus, the finding that technological change leads to "upsizing" may be due to the nature of the innovations represented in the SPRU file, which consist mainly of successful product innovations. This implies that Van Reenen's findings may simply reflect evidence of high private returns to investment in R&D. Still, his study does underscore the importance of including controls in the model for industry and timing effects.

In Table 5.4, I control for industry and timing effects, as outlined in Eq. 5 on page 37. Again, the results are consistent with the hypothesis that compositional changes favor highly educated workers. Although not all of the changes in shares are statistically significant, the results suggest that technology adoption stimulates an increase in the demand for each class of nonproduction worker and a sharp decline in the demand for production labor (both the level and the share). Thus, it appears that the evidence is consistent with Hypothesis 3: AMT is associated with downsizing and skill upgrading.

Two-stage probit estimates of the share equations are also presented in Table 5.5. These regressions include the proportion of sales to the government as an additional control variable. (Recall that the purpose of using this estimation procedure is to control for the endogeneity of technical change.) As I have shown, both the unconditional and conditional probabilities of AMT adoption are strongly related to a set of firm characteristics.

As noted in Chapter Two, the efficiency of the two-stage estimates depends critically on the "fit" in the first stage of the model and on the assumption that the instruments are uncorrelated with the error term in the second stage. Our use of the two-stage procedure is supported by the highly significant F-tests in the first stage of the model. Also, the correlations between the instruments and the residuals from the second stage of the model are uniformly insignificant. It is interesting to note that the findings are virtually the same as those reported in Table 5.4; that is, the two-stage results also suggest that technology adoption leads to a shift in labor composition in favor of nonproduction workers.

In Chapter Four, I hypothesized that it is important to distinguish between linked and integrated AMTs. Furthermore, I argued that linked and integrated AMTs should have different compositional and empowerment effects, because linked technologies are generally asso-

Table 5.5 Two-Stage Probit Estimation of Employment Equations (estimates of β_1 from Eq. 5)[a]

	MS[b]	TP	R&D	CL	DPL	OTH
Employment share equations						
Year $t+1$	0.059***[c]	0.063	0.058	-0.046	-0.073***	-0.053***
	(1.98)[d]	(1.73)	(1.47)	(1.58)	(2.16)	(2.15)
Year $t+2$	0.111***	0.090**	0.063	-0.059	-0.137***	-0.064***
	(2.16)	(1.97)	(1.51)	(1.73)	(3.16)	(2.25)
Year $t+3$	0.128***	0.110**	0.092	-0.044	-0.128***	-0.075***
	(2.24)	(2.01)	(1.95)	(1.37)	(3.06)	(2.11)
Employment cost share equations						
Year $t+1$	0.084	0.073	0.058	-0.047	-0.071***	-0.052***
	(1.92)	(1.93)	(1.49)	(1.62)	(2.14)	(2.33)
Year $t+2$	0.096	0.088	0.068	-0.058	-0.138***	-0.067***
	(1.94)	(1.94)	(1.67)	(1.74)	(3.18)	(2.58)
Year $t+3$	0.108***	0.101**	0.079	-0.042	-0.128***	-0.072**
	(2.23)	(2.04)	(1.79)	(1.33)	(3.04)	(1.96)

[a] Each regression includes an industry dummy.
[b] MS = managerial and supervisory.
TP = technical and professional.
R&D= research and development.
CL = clerical and administrative.
DPL = direct production labor and supporting personnel.
OTH= other production labor.
[c] *** = statistical significance at the 1% level.
** = statistical significance at the 5% level.
[d] t-Statistics are in parentheses.

ciated more with quality enhancement and integrated technologies are more closely associated with the displacement of labor.

To determine whether these categories for AMTs are valid, I performed a factor analysis, a method that has been used extensively in psychology and sociology when scholars assess variables such as intelligence, prestige, or status, which are difficult to measure precisely and have multiple indicators.[1] The idea behind factor analysis is that there is a hypothetical, unobservable variable that a set of variables share in common: that is, if the observables share a strong positive correlation, then it is assumed that they can also be categorized as capturing a measure of a common unobservable "factor."

For this book, I propose that CAD, CAM, CNC, SPC, and JIT are all linked advanced manufacturing technologies, and that FMS, AS/R, AGV, ROB, and GT are integrated advanced manufacturing technologies. To test these hypotheses, we compute the intercorrelations for all types of AMTs and extract the general factor common to them all.

I use the standard factor analysis method, with principal components being the extraction procedure and VARIMAX rotation. Three factors with eigen values greater than 1 were extracted, which accounted for a total of 58% of the common variance. Table 5.6 presents the rotated factor structure matrix for 12 advanced manufacturing technologies.

Our results indicate that AMT adoptions do indeed factor into two dimensions, linked and integrated AMTs. Factor 1 consists of what was previously described as linked AMT. Marker items (items with loadings greater than 0.50) are CAD, CAM, CNC, and SPC.[2] Factor 2 consists of what was previously described as integrated AMT. Marker items include AAS, AS/R, FMS, ROB, and GT.

Several AMTs did not load on the expected factor. Inconsistent with expectations, AGV did not load on the second factor, and JIT and MRP did not load on the first factor, but instead loaded on the third. While JIT and MRP might be beneficial technologies for waste reduction and production planning, these technologies may not be prerequisites or essential components of linked or integrated AMTs. This result may also be explained by the job-shop nature of many of the firms, because JIT is better suited for assembly line operations.[3]

I chose to limit further analyses to the first two factors, because they are clearly in line with the distinction previously outlined between

Table 5.6 Rotated Factor Structure Matrix of AMTs

AMT	Factor 1	Factor 2	Factor 3
AAS (automated assembly system)[a]	0.1066	0.7316	0.0507
AGV (automated guided vehicle)	0.1321	0.2437	0.3269
AS/R (automated storage and retrieval)	0.1712	0.5365	0.1777
CAD (computer-aided design)	0.7965	0.2724	0.1563
CAM (computer-aided manufacturing)	0.7834	0.2226	0.0417
CNC (computer numerical control)	0.8829	0.0461	0.0218
FMS (flexible manufacturing system)	0.1555	0.6803	0.1480
GT (group technology)	0.1279	0.6547	0.2743
JIT (just-in-time production)	0.0001	0.0581	0.8035
MRP (materials requirements planning)[a]	0.1141	0.1191	0.7300
ROB (robotic device)	0.0845	0.6932	0.3583
SPC (statistical process control)	0.6660	0.1210	0.3992

[a] AAS and MRP do not fit into either the linked or integrated categories.

linked and integrated AMTs. The bottom line is that the results provide strong empirical support for Hypothesis 5. For subsequent analyses, I constructed two dummy variables for linked and integrated AMT based on the loading factors just described. A total of 61 firms adopted at least one linked AMT, while 25 adopted at least one integrated AMT.[4]

I also hypothesize that the adoption of AMTs will lead to a reduction in the workforce, accompanied by increased technical skill levels to deal with increased technological complexity. This will result in a workforce that favors technical/professional, managerial, and R&D workers. However, I also conjecture that the nature of these recompositional shifts will differ by category of AMT. Given that integrated AMTs embody a more "advanced" state of technology (requiring more scientific and engineering input) and that they automate materials handling and control (which previously required many low-skilled employees), I expect that greater reductions in low-skilled staff should result from integrated AMTs.

The top panel of Table 5.7, shows the percentage of companies that employed methods of implementation that are likely to lead to an

Table 5.7 Firms Adopting Methods of AMT Implementation That Are Consistent with an Enhancement in Employee Empowerment

Methods of implementation	Linked AMT firms	Integrated AMT firms	Difference
Percentage of firms reporting			
Training existing personnel	91.8	62.5	N/A
Changing employees' job responsibilities	50.5	25.8	N/A
Create new jobs and career opportunities for employees	29.6	11.7	N/A
Increased employee control	40.7	24.5	N/A
Mean values of empowerment measures[a]			
Training existing personnel	0.324	0.073	0.251***[b] (6.22)[c]
Changing employees' job responsibilities	0.183	0.037	0.146*** (4.90)
Create new jobs and career opportunities for employees	0.099	0.011	0.088*** (3.92)
Increased employee control	0.287	0.104	0.183*** (5.71)

[a] Scores can range from 0 to 1.
[b] *** = statistical significance at the 1% level.
[c] t-Statistics are in parentheses.

enhancement in employee empowerment. Approximately 92% of the firms that adopted at least one linked AMT reported that implementation involved training existing personnel. About half of the companies that adopted a linked AMT also reported that implementation led to changes in employees' job responsibilities. Note that for all four indicators, the percentages are higher for linked AMT firms than for integrated AMT firms. Based on these crude measures, it appears that implementation of linked AMTs may be more closely associated with the enhancement of empowerment than that of integrated AMTs.

To formally test this hypothesis, I computed scores for methods of implementation that I interpret as indicators of employment empowerment. These scores are summed and divided by the number of linked or integrated AMTs that the firm adopted (at most, four for linked

AMT firms and five for integrated AMT firms) so that all scores are on the same dummy-coded, 0-to-1 scale. For example, let us assume that a company has adopted all four linked technologies. If this firm checked CNC and SPC with regard to "training existing personnel" (see the Appendix) but left CAD and CAM blank, the firm would get a score of 0.5 (2/4) for this empowerment indicator.

In the bottom section of Table 5.7 are the mean empowerment scores for linked AMT firms and integrated AMT firms, and in the third column, the difference in the mean values and the t-statistic. All four of the empowerment measures are lower for integrated AMT firms, confirming that the incidence of methods that promote empowerment is significantly greater for linked than for integrated AMTs. All of the differences in column 3 are statistically significant at the 1% level.

Table 5.8 shows the downsizing and compositional effects for each class of technology and for all AMTs. The first three columns present the mean and median percentage changes in total employment, and the remaining columns show the changes in employment shares for six categories of workers over the sample period for firms that adopted new technologies. The empirical findings are consistent with Hypotheses 6 and 7, (pp. 53 and 55); that is, our findings suggest that the association between downsizing and integrated AMTs is stronger than the association between downsizing and linked AMTs. Furthermore, it appears that the "skill bias" of technological change or the recompositional effects in favor of highly educated workers is greater for integrated AMTs than for linked AMTs.

These results must be interpreted with caution because they do not include industry controls. Many companies on Long Island are in defense or defense-related industries. The leading prime defense contractors in the region are Grumman (now part of Northrup Grumman), Paramax Systems, AIL Systems, Militope, and Loral Corporation, and many smaller firms on Long Island rely heavily on subcontracting from firms such as Boeing and McDonnell Douglas.

Kamer (1993) reported that defense-related jobs accounted for 45.1% of total manufacturing employment in 1985 (the peak of the defense buildup during the Reagan years) and approximately 8% of total nonfarm employment. During the late 1980s, many Long Island companies experienced a dramatic downturn in demand as the Cold

Table 5.8 Mean and Median Employment Levels, Percentage Changes, and Shares for AMT Firms[a]

| | Total employment | | | Mean (median) employment shares (%) | | | | | | | | | | | |
| | Mean (median) | | Mean (median) change | MS[b] | | TP | | R&D | | CL | | DPL | | OTH | |
Number of AMTs used	1987	1990	(%)	1987	1990	1987	1990	1987	1990	1987	1990	1987	1990	1987	1990
Any AMT	777	589	−5.8	10.1	11.2	8.9	14.1	6.2	6.7	16.2	13.5	62.8	60.0	2.1	1.3
	(189)[c]	(160)	(−7.1)	(9.5)	(10.5)	(9.5)	(13.1)	(2.9)	(3.3)	(16.0)	(12.3)	(66.3)	(66.7)	(0.0)	(0.0)
Linked AMT	402	328	−3.2	12.5	13.7	8.6	10.4	7.5	7.2	14.8	15.7	65.4	63.8	3.1	2.1
	(113)	(104)	(−3.9)	(10.9)	(11.3)	(9.2)	(9.9)	(3.5)	(3.4)	(14.4)	(14.8)	(67.4)	(67.2)	(0.0)	(0.0)
Integrated AMT	1025	779	−9.4	7.7	8.7	9.2	17.8	4.9	6.2	17.8	11.3	60.2	56.2	1.1	0.5
	(323)	(256)	(−10.3)	(7.3)	(8.2)	(9.9)	(15.3)	(2.4)	(3.8)	(16.3)	(11.7)	(65.3)	(63.7)	(0.0)	(0.0)

[a] No industry controls.
[b] MS = managerial and supervisory.
TP = technical and professional.
R&D = research and development.
CL = clerical and administrative.
DPL = direct labor and supporting personnel.
OTH = other production labor.
[c] Medians are in parentheses.

War ended: between 1985 and 1991, Long Island lost over 29,000 jobs in defense-related industries (approximately 36% of all its defense-related jobs). Therefore, some of the downsizing and compositional changes could be due to a decline in industry demand. To account for this decline, I included the proportion of sales to the military as an additional regressor. Furthermore, an analysis of changes over the sample period for technology adopters does <u>not</u> control for the <u>timing</u> effects of the technology adoption.

I controlled for industry and timing effects by estimating the following set of equations:

Eq. 9 $\ln (s_{ijt+2}/s_{ijt-1}) = \alpha_j + \beta\, AMT_t + \gamma\, GOVSALES_i$

Eq. 10 $\ln (s_{ijt+2}/s_{ijt-1}) = \alpha_j + \beta\, LINKAMT_t + \gamma\, GOVSALES_i$

Eq. 11 $\ln (s_{ijt+2}/s_{ijt-1}) = \alpha_j + \beta\, INTAMT_t + \gamma\, GOVSALES_i$

where the dependent variable in each equation is the growth rate (logarithmic change) in the employment share (s_{ij}) for firm i in industry j; AMT_t equals 1 if the firm adopted an advanced manufacturing technology between $t-1$ and t and equals 0 otherwise; $LINKAMT_t$ equals 1 if the firm adopted a linked AMT between $t-1$ and t and equals 0 otherwise; $INTAMT_t$ equals 1 if the firm adopted an integrated AMT between $t-1$ and t and equals 0 otherwise; and GOVSALES is the firm's percentage of sales to the government. Each of these growth rates is standardized by industry.

The coefficient (β) on the dummy variable will provide an estimate of the difference in the growth rate of the share between technology adopters and non-adopters. Note also that there are now two dummy variables for AMT, so I can examine employment shifts for linked and integrated AMT separately. This proposed specification also enables us to examine the timing of labor force adjustments, because we specifically examine compositional changes after the technology adoption.

The values of ß from these regressions are reported in Table 5.9. I estimated 18 regressions: 3 technological classifications (all AMTs, linked AMTs, and integrated AMTs) × 6 classes of workers. The point estimates can be interpreted as follows: the value 0.131 in row 3, column 1 implies that the share of managerial and supervisory employees

Table 5.9 Differences in Mean Growth Rates of Employment Shares between Firms Adopting AMTs in Year t versus Those Not Adopting AMTs[a]

Type of AMT adoption	MS[b]	TP	R&D	CL	DPL	OTH
Overall	0.107***[c]	0.103***	0.056	−0.051	−0.161***	−0.086***
	(2.09)[d]	(2.11)	(1.42)	(1.71)	(3.52)	(3.01)
Linked AMT	0.083*	0.088**	0.046	−0.036	−0.124***	−0.065***
	(1.82)	(1.94)	(1.25)	(1.44)	(2.32)	(2.63)
Integrated AMT	0.131***	0.121***	0.089*	−0.081	−0.184***	−0.099***
	(2.77)	(2.34)	(1.90)	(1.83)	(4.01)	(3.21)

[a] Estimates of β from Eq. 9, 10, and 11.
[b] MS =managerial and supervisory.
TP =technical and professional.
R&D=research and development.
CL =clerical and administrative.
DPL =direct labor and supporting personnel.
OTH =other production labor.
[c] *** = statistical significance at the 1% level.
** = statistical significance at the 5% level.
* = statistical significance at the 10% level.
[d] t-Statistics are in parentheses.

(fraction of the workforce devoted to this type of worker) increases by 13.1% in firms that adopt integrated AMTs, relative to firms in the same industry that do not adopt integrated AMTs. The mean difference in the growth rate of this share is highly statistically significant.

Similar increases in the share of technical and professional employees and of scientists and engineers are found for integrated AMTs (increases of 12.1% and 8.9%, respectively). There is a relative decline of about 18% in the share of production workers for firms that adopt integrated AMTs. While the compositional shift towards workers with higher levels of education and skill (positive shifts in columns 1–3, negative shifts in columns 4–6), is evident for all AMTs and linked AMTs, the differences (between adopters and non-adopters) are uniformly of high statistical significance only in the case of integrated AMTs. Also, the positive shift in the proportion of scientists and engineers is significant for integrated AMTs, while insignificant for linked AMTs. These findings are consistent with the hypotheses outlined in Chapter 4.

These findings raise several issues which are beyond the scope of this study. First, it would be interesting to untangle the causal relationship between technological change and changes in labor composition. In my framework, I implicitly assume that the adoption of a new technology leads to changes in workforce composition. However, both types of decisions may be jointly determined. For example, the decisions to restructure the workforce and to implement a technological change may occur concomitantly, rather than one causing the other. Also, there may be feedback effects associated with an increase in R&D employment. For example, several authors have argued that an increase the number of scientists and engineers can be viewed as a source of information acquisition regarding the existence and successful implementation of new technologies (e.g., see McCardle 1985). In future research, I plan to explore some of these linkages.

In my framework, I also implicitly assume that technological change is the sole determinant of changes in labor composition. The growth in foreign trade has also been cited as a possible cause of the decline in demand for low-skilled workers. In a recent study based on industry-level data, detailed measures of the educational attainment of the workforce, and a dynamic cost function framework, Catherine Morrison and I (Morrison and Siegel 1996) examined the simultaneous effects of trade, technology, and outsourcing on labor composition. We found that technological factors (such as investment in R&D and computers) had the strongest effect on shifts in favor of highly educated workers. Trade had a very small direct effect, which is consistent with the findings of most recent studies (Lawrence and Slaughter 1993).[5] We did find, however, that trade may have an indirect effect because it stimulated additional investment in technology, which in turn leads to changes in labor composition.

Several important caveats to the findings presented in this chapter must be noted. First, the findings are based on a regional sample that is weighted towards defense-related industries and thus may not be representative of national trends. Also, my focus is on manufacturing firms and technologies. It would also be interesting to explore the effects of technological change on labor composition and empowerment in service industries, since services constitute a large and growing fraction of employment and output.

In this chapter, I have presented evidence on the antecedents and labor market outcomes of technology adoption. Technology usage appears to be positively related to firm size, R&D intensity (a form of "information acquisition"), and cumulative experience with related technologies. Technology usage also appears to be associated with corporate downsizing, skill-upgrading, and employee empowerment.

I also found evidence supporting for the hypothesis that we can distinguish between linked and integrated AMTs. This may be critical in the consideration of "skill-biased" technological change, because there appear to be fundamental differences in the labor market outcomes associated with the adoption of these two classes of AMTs. Linked AMTs appear to be more closely associated with efforts to enhance employee empowerment, while integrated AMTs appear to be more closely associated with downsizing and shifts in labor toward of highly educated workers. Thus, it appears the skill bias of technological change may be considerably stronger for more advanced, integrated technologies.

Notes

1. Factor analysis was invented by Spearman (1904), who used it in his attempts to measure intelligence (what he termed a g [general intelligence] factor). See Gould (1981) for an excellent, nontechnical history of factor analysis.
2. A "loading" is a measure of the correlation between an item and an unobserved common "factor." A marker item is an item (in this case, an AMT) that is highly positively correlated with a given factor (usually defined as >0.50) and does not "load" on any other factor. (See Gorsuch [1974] for details).
3. This is an observation that was culled from my field research, involving factory visits to 20 companies. These visits clearly revealed that for many of these firms, the production process is of a "job shop" nature.
4. In our sample, there are 18 companies that adopted both linked and integrated AMTs. While it may be true that there is something special about firms that use both linked and integrated AMTs, I believe that I need not exclude these companies or analyze them separately. Thus, the empirical results include these overlapping firms.
5. A notable exception is Revenga (1992), who reported a strong trade effect.

6 Case Studies

This chapter summarizes case studies of 4 out of 20 firms I visited that had completed the survey and whose managers and other employees were willing to discuss their experiences with AMT investment. My objective was to have company officials elaborate on certain points that could not be addressed in the survey, generally qualitative information on the costs and benefits of AMT investment and changes in the work environment. I also thought it would be useful to observe how technologies were implemented on the factory floor and in other functional areas of the business.

I selected the four companies for an in-depth, case-study analysis because these firms are in different industries and because managers and workers were willing to meet with us and provide detailed information on sensitive topics. These four companies vary greatly in terms of size, scope, and level of technology usage. While there is obviously a degree of sample selection bias inherent to this approach, the cases may provide useful information for policymakers and academics who wish to assess the economic and managerial implications of technological change.[1]

SYMBOL TECHNOLOGIES

Company Background

Symbol Technologies is the world's largest manufacturer of bar-code-scanning devices (data capture systems), with nearly two million scanners and terminals installed. Bar codes consist of a series of lines or bars printed on a contrasting background on packages and products. Symbol's devices read data from the bar code, which usually includes information about the item such as location, cost, price, and manufacturer. Bar-code scanning greatly simplifies data entry and enhances productivity in a wide range of sectors, including supermarkets, retail establishments, large department stores, manufacturing, package and parcel delivery, warehousing and distribution, health care, and many other industries.

By using these devices, companies are able to track sales and inventories quickly and efficiently, promote security, and reduce employee theft. In recent years, sales of portable devices have increased substantially as advances in computer technology have enabled Symbol to produce smaller scanners at relatively moderate prices.

Symbol's competitive strategy is differentiation (the firm offers a wide range of products varying in price and quality), with a strong emphasis on quality and innovation. A major source of their competitive advantage is their technological leadership. To maintain this edge, they invest heavily in product development. The company's manufacturing strategy is consistent with the competitive strategy. They produce many customized products and their manufacturing facilities embody the latest technological developments. The implementation of AMTs has made Symbol more responsive to customer needs, by enabling the firm to endow its products with features that are most desired by key customers, such as large supermarket and retail establishments and overnight delivery companies. Rapid response to customer needs is also a critical success factor in this industry.

Like many of the firms I visited, Symbol faces increasing competitive pressures. First, the product life cycle has been shortened dramatically, in part due to the widespread use of AMT. The use of CAD/CAM, in particular, has enabled the firm to introduce new products quite rapidly. Second, although Symbol has substantial brand loyalty, several foreign competitors have captured a growing share of the market in recent years. During our analysis of the company, the firm was in the process of consolidating its major manufacturing facilities. A new state-of-the-art facility on Long Island will combine manufacturing operations from Long Island and California. After the construction of the new plant, corporate headquarters, R&D, and manufacturing plants will all be located in Suffolk County at the Bohemia, New York, facility.

Findings Based on Interviews and Plant Visits

With three graduate students, I visited with Mr. William Dowlin, Vice President of Operations for the Bar Code Scanning Division, who reported that Symbol is an active and enthusiastic user of AMTs. They have successfully implemented 7 of the 12 AMTs: computer-aided design (CAD), computer-aided manufacturing (CAM), automated storage and retrieval (AS/R), flexible manufacturing system (FMS), materials requirements planning (MRP), statistical processing control (SPC), and robotics (ROB). The company's newest CAD/CAM system has enabled them to design and introduce innovative products. Their newest hand-held scanner, the LRT 3800, combines wireless radio-frequency data communications technology with laser-based bar-code scanning. It is now widely used in retail establishments.

The major benefits of AMT adoption have been more rapid product development, reductions in labor cost, better reliability, and especially, flexibility. Symbol has clearly taken the concept of flexible manufacturing seriously, by switching from high volume to differentiation and small volume. Buyers want customized products that are based on state-of-the-art computer technologies. The use of AMT allows Symbol to deliver these products at a reasonable price. This has enhanced its corporate strategy and increased profitability.

The firm has invested heavily in automation, which it believes has increased quality and flexibility. As automation has proceeded, Symbol has strived to upgrade workers' skills. The human resource management policies of the company actively promote skill upgrading through an emphasis on training and education. Mr. Dowlin mentioned the importance of local universities in the firm's efforts to upgrade the skills of its workers (such as the engineering school at SUNY-Stony Brook, Polytechnic University, and others). The workers we spoke to on the factory floor were exposed to many training seminars.

The workers also discussed another important change in the work environment that resulted from implementation of the new technologies: a greater emphasis on teamwork and consensus. Employee participation in managerial decision making also increased. Joint consultation committees were formed to provide additional outlets for worker feedback and communication. The bottom line is that the changes in work environment appear to be part of an overall strategy to raise the level of worker commitment to the firm's overall goals and objectives. Based on broad measures of changes in worker productivity and firm profitability, that strategy appears to have been successful.

Mr. Dowlin reported that another important change induced by AMT investment was an increase in outsourcing. The firm now outsources many low-skilled, labor-intensive functions. This has enabled the firm to focus its efforts on improving product quality and innovation. While reducing the number of unskilled workers, it has increased the number of technicians. Several of the technicians we spoke to were originally hired as unskilled workers.

Like many of the firms I visited, Symbol is rather undisciplined in deriving rate of return estimates for R&D projects. Although it usually doesn't generate formal estimates of actual or expected returns, Mr. Dowlin surmises that there have been substantial labor cost savings associated with AMT investment, something on the order of about 5% per year (which is consistent with our best estimates, based on the survey response).

According to Mr. Dowlin and many of the workers I interviewed, the major benefit of AMT investment was greater integration across functional activities. Most of the advanced manufacturing technologies, particularly those that we have labeled linked AMTs, allow workers to exchange information across functional areas, such as R&D, production, finance and accounting, logistics, purchasing, and marketing.

These technological investments have enabled Symbol to greatly reduce its clerical staff and middle management. Contrary to general trends, however, it has not reduced the number of production workers. The largest employment increase is in the number of technical and professional employees. Mr. Dowlin implied that this increase does not reflect new hires, but rather the upward mobility of existing workers, which is consistent with the firm's stated objective of upgrading the skills of workers. Not surprisingly, Symbol had one of the highest empowerment scores in our sample.

Mr. Dowlin also asserted that AMT has changed Symbol's relationships with suppliers. Since it began implementing AMTs, it has fewer suppliers but has closer relationships with them. It has actively encouraged its suppliers to undertake investments in AMT.

Symbol has also benefited from its suppliers' investments in AMT. These firms have reduced their labor costs and increased quality and flexibility. For example, Symbol's hand-held scanners, its best-selling products, are made of plastic. Long Island is a hub for manufacturers of molded plastic products. Many of these plastics firms have invested in programmable machinery that has significantly improved their productivity and reliability, enabling them to maintain constant prices. Symbol's scanners also contain computer chips, which have fallen dramatically in price. Mr. Dowlin noted that because of these trends, Symbol's materials prices have fallen over the past few years. This accounts for some of the rise in Symbol's profitability.

LUMEX CORPORATION

Company Background

Lumex Corporation designs and manufactures exercise, rehabilitation, and health care equipment. Its Cybex division produces exercise and fitness equipment. Cybex equipment is used by almost all professional sports teams and has gained wide acceptance among doctors for use by orthopedic and neurological patients. The company's line of "isokinetic" products, in which resistance accommodates the amount of force applied throughout the complete range of motion, is quite popular with doctors because resistance stops or decreases automatically if the patient feels pain or fatigue. Cybex has emerged as the leader in sales to health clubs, with a market share of approximately 15%.

The firm's Lumex division manufactures products that are geared toward geriatric care and patient aids. These include walkers, canes, crutches, and commodes, as well as patient seating, over-bed tables, etc. These products are used in home health care and institutional markets. Obviously, demographic trends are quite favorable in this sector. In 1990, the division launched the most successful new product in the company's history, the Cybex model 6000, a new generation of extremity testing and rehabilitation equipment. This product confirmed Cybex's technical leadership in the industry.

The Cybex and Lumex divisions contribute almost equally to over-all revenue. Described by securities analysts as a classic, small, growth company in the 1980s, Lumex lost market share when the patents for several key products expired, and it diverted resources to the develop-ment of machines and equipment for back rehabilitation. The core business suffered from changes in Medicare and Medicaid reimburse-ment along with greater price competition. Given its competitive strat-egy of focused differentiation (the firm offers a small range of higher-priced, high-quality products) and thus a strong emphasis on quality, the firm has invested heavily in AMTs.

Findings Based on Interviews and Plant Visits

I interviewed Mr. Russell Olesen, Vice President of Manufactur-ing, and conducted several plant visits with two graduate students. In contrast to most of the companies I visited, Lumex generates detailed projections of returns on investment in new technologies. Mr. Olesen mentioned that the emphasis on numbers was an important aspect of the corporate culture, because the leaders of the company have tradi-tionally had an accounting background. Furthermore, Lumex actually conducts an *ex post facto* analysis of rates of returns; that is, several years after implementation, it assesses whether the targets have been met. If these targets have not been reached, it tries to identify the sources of inefficiency. Mr. Olesen provided me with detailed reports documenting the cost savings, along many dimensions, from AMT investment.

When I visited Lumex, the firm was in the process of implement-ing a just-in-time inventory system. It was especially concerned with reducing the size of its warehouse. Not surprisingly, it expected JIT to lead to changes in job responsibilities and increased managerial con-trol. Lumex's products embody the latest features, such as computers, and are priced accordingly. A major source of its competitive advan-tage is its ability to generate new products and improve existing ones. Consistent with its corporate-level strategy, it devotes a fairly large proportion of revenue to R&D.

The strong emphasis on product innovation is reflected in the num-bers provided us. Lumex devotes about 6% of revenue to R&D invest-ment. Furthermore, over the sample period (1987–1990), it reported

an increase of 67% in the number of scientists and engineers. Mr. Olesen agreed with the notion that hiring more R&D staff served as a form
of knowledge acquisition, since these new hires are generally quite
familiar with the nuances of AMT investment.

Our discussions with Mr. Olesen and other employees revealed
that AMT adoption led to several key changes in human resource management policies. First, there was some downsizing when the technologies were first implemented. Second, the company devoted more
resources to training the overwhelming majority of employees who
remained with the firm. Mr. Olesen pointed out that the company's
policy was to train all workers who could potentially encounter the
new equipment, including maintenance employees. This was viewed
as a way to safeguard the firm's substantial financial investment in
AMT. That is, the introduction of programmable machinery involved
the use of very precise and expensive machines and processes. This, in
turn, required more complicated work in performing proper maintenance of this equipment and more sophisticated analysis to ensure efficient utilization.

A third change in HRM policy in the aftermath of AMT adoption
involved modifications in the nature of work and skill requirements.
Several employees mentioned that their supervisors had given them
more decision-making authority. They were also asked to perform
additional clerical tasks, such as processing and analyzing data. Some
workers mentioned that they had developed managerial skills, which
they had acquired by attending the local community college. Finally,
interpersonal skills were increasingly important, as there was a much
greater emphasis on teamwork. In recent years, the firm had established cross-functional teams from engineering, production, procurement, and customer support to take advantage of the benefits of
integration that resulted from AMT adoption.

According to Mr. Olesen, the primary difficulty in implementing
the new technologies was "tailoring a generic system to meet the special needs of our firm." He pointed out that by definition, a software
program that controls a CNC machine and handles materials which are
used in the production process in a flexible manufacturing system must
be unique to that particular product line (in this case, rehabilitative
equipment) and customer needs. Most of the additional expense associated with AMT implementation was devoted to hiring consultants to

customize the software and to train workers to use the new software. He cautioned managers who are contemplating similar investments to factor these adjustment costs into the projections of returns on investment in new technologies.

PALL EAST HILLS MANUFACTURING

Company Background

Pall Corporation is the world's leading manufacturer of filtration devices. Its products are used to remove microscopic contaminants of solids, liquids, or gases. There are three markets for these filtration devices: healthcare, aerospace, and fluid processing. Health care is the fastest-growing segment, although fluid processing is also doing well because of the substantial increase in environmental regulations. Most of Pall's products are proprietary filter media produced by melt-blowing of polymer fibers, chemical film casting, papermaking, and metallurgical processes.

Health care products constitute approximately 50% of sales. These products include filters to protect patients against contamination from intravenous fluids, transfused blood and blood components, and breathing gases. Filters are also used in diagnostic devices to assess diseases in plants, animals, and humans; to generate sterile, contamination-free pharmaceuticals, biopharmaceuticals, and biologicals; and to produce yeast-free and bacteria-free water, beverages, and food products. The health care market also includes a range of electronic test instruments that enable users to easily test the integrity of their filters before and after use.

The aerospace market represents 30% of sales. The firm provides a broad line of lube oil, fuel, and air filters containing proprietary filter media. This group also produces filter manifolds, self-cleaning engine intake filters, and pressure swing absorption (PSA) systems. For users of fluid-power or lubricated machinery, Pall provides filter elements and housings to control particulate contamination in hydraulic or lubricating oil. In the mobile fluid-power market, Pall provides transmission fluid and diesel engine filters.

The fluid processing market, which represents approximately 22% of sales, includes products that remove contaminants from liquids and gases in process streams through the retention and recirculation of catalysts and with the minimization of hazardous waste. Growth in environmental regulation and awareness has stimulated demand.

Findings Based on Interviews and Plant Visits

Pall pursues a focus strategy, stressing quality, not price. It has invested heavily in new manufacturing technologies, with a strong emphasis on increasing the effectiveness of materials and inventory management. Although the firm is innovative, it does not seek "first mover" advantages in undertaking technological investments. Pall's reluctance to be the first to implement a new technology is consistent with its corporate culture, which is very conservative and risk averse. This conservative policy may stem from some negative experiences with previous AMT investments.

Accompanied by two graduate students, I interviewed Mr. Jack Caulfield, the Vice President of Manufacturing. Mr. Caulfield and other company officials were concerned about the disruptive nature of technological change. They mentioned several times that they wanted the process to be "evolutionary," not "revolutionary." Many of the companies we visited had complaints regarding the quality of software and difficulties with compatibility. Mr. Caulfield is confident that his organization has learned from these experiences. He mentioned that recent AMT investments had yielded higher rates of return than previous ones, implying that significant learning effects had been associated with successive AMT implementations. He was still struggling to transmit this message to several leading corporate executives, who were disappointed with past financial results.

Instead of advancing a learning argument, Mr. Caulfield was promoting the use of new technologies as a means of gaining additional control over inventory, work orders, and other operational aspects. Mr. Caulfield also stressed how AMTs could be used to enhance accounting controls. Now that it had achieved inventory controls, Pall wanted to achieve greater control from an accounting standpoint. The company plans to achieve this through computerization, based on job costing information provided by data stations (called "elves") that read workers' badges, automatically identifying the operation they perform and automatically recording the date and time. The close of the operation would also be recorded. This process is called job costing and workflow analysis. Pall also proposes to use bar-coding technology to improve the accuracy of data that is used to assess productivity.

Mr. Caulfield emphasized that Pall has not used AMT specifically to downsize the labor force. Its objective is to raise productivity while maintaining stability of the workforce. The reported numbers are consistent with this. Pall stressed the importance of using the technology to empower workers and to increase their awareness of activities taking place in other functional areas. The only area of the business where AMT is likely to lead to workforce reductions is in the area of processing purchase orders. Like many other companies on Long Island, Pall experienced some difficulty in implementing the software to achieve this goal. When we interviewed Mr. Caulfield and other company officials, they were attempting to overcome these difficulties, which would result in streamlining the purchasing function.

In contrast to most of the firms we visited, Pall was more focused on upgrading the skills of its engineers. Mr. Caulfield pointed out that some of his engineers graduated more than a decade ago, when it was impossible to work with such sophisticated computers and machinery, much less a whole computer-integrated manufacturing system. Given the critical role that engineers play in design, implementation, problem solving, and continuous improvement of computer-based technologies, they must periodically be retrained. In this regard, Pall has taken advantage of several university-sponsored educational programs.

Mr. Caulfield pointed out that middle managers must also be retrained, both attitudinally and technically, to cope with new technologies and the greater emphasis on integration, a major theme of most AMTs. He also asserted that it sometimes is difficult for managers to adjust to changes in the organizational environment and a certain "loss of control" that results from AMT usage. Many experienced managers were especially uncomfortable with what they perceived to be a loss of authority.

Other major organizational changes were a greater emphasis on teamwork and much greater flexibility in work assignments. The workers we interviewed also noted that new technologies were associated with a move towards multi-skill and broader-scope jobs. They also mentioned that they were now much more involved in planning and control, which suggests that the firm was engaging in job "enrichment" (see Heizer and Render 1999), another aspect of employee empowerment. These reflections led Mr. Caulfield to admit that the company had still not sorted out the full implications of these changes in human resource management policies, since there was still some resistance to change among "old guard" managers (those with the firm before these changes were implemented).

ELECTRONIC HARDWARE

Company Background

Electronic Hardware Corporation, located in Farmingdale, Long Island, is a leading manufacturer of control knobs, switches, and other panel hardware for military, industrial, and commercial equipment. Their primary customers are Grumman, Boeing Aerospace, the Army, and the Navy.

The firm has an established reputation as a manufacturer of high quality and reliability, and it has only one major competitor. Currently, 50–60% of its sales is derived from government/military contracts. Despite the general downturn in demand from the military sector, its annual sales revenue has more than doubled between 1987 and 1994. This can be attributed mainly to three recent acquisitions of smaller competitors.

Although they have only 78 employees (including a large number of part-time workers), Electronic Hardware is now producing over 10,000 type of knobs and control devices. This has helped them capture market share from competitors who can offer only a limited range of products. Unfortunately, the large variety of products and the recent consolidations have also created severe problems in the coordination of purchasing, planning, inventory management, and especially with processing customer orders.

Electronic Hardware is also in the process of implementing a major change in corporate strategy. The catalyst for this strategic initiative is the recent downturn in demand in the military sector. This has compelled the firm to shift its focus from defense industries to commercial and industrial applications. To achieve this strategic goal, Electronic Hardware is an active participant in the New York State Diversification Program, which is helping companies reduce their dependence on the defense sector.[2] They are also participating in the New York State Industrial Effectiveness Program, a total quality management (TQM) program partially financed by New York State and the Long Island Lighting Company (LILCO). The Industrial Effectiveness Program also provides subsidies for training and technology, as well as rebates on electric rates.

The firm has not yet implemented an AMT, but they are interested in implementing material requirements planning and just-in-time inventory systems. They realize that the success of their efforts to diversify depends critically on their ability to respond to customer needs. The investment in AMT will be used to provide flexibility and enhance efficiency, which are essential to achieving and sustaining a competitive advantage in these new industries. Furthermore, it seems likely that competition, especially technological competition, will be more intense in commercial and industrial markets than in defense industries.

Findings Based on Interviews and Plant Visits

I interviewed Mr. Stephen Sgammato, the Controller of Electronic Hardware, on several occasions and conducted several plant visits accompanied by two graduate students. At the time, the firm was in the process of selecting a computer-integrated manufacturing (CIM) system. Mr. Sgammato noted that a major competitor has fully automated production capabilities and has computerized almost all of its administrative functions. He recognizes the need for technological advancement. Our discussions focused primarily on the barriers to AMT implementation and how they can be surmounted.

It was clear from the plant visits that there is a strong need for a CIM system. There is a substantial amount of excess inventory at the manufacturing facility. Both the warehouse and the administrative offices were cluttered, and the firm's existing hardware and software were inadequate to process the large volume of orders. Currently, most of the simple office work, such as bill of materials, routing, scheduling, and costing, is done manually. The current computer system, an IBM 36 model, is used primarily for administrative purposes.

Mr. Sgammato cited two major obstacles to the adoption of AMTs. The first is difficulty in quantifying the benefits of implementing AMTs. He noted that this was a major factor in their decision not to adopt computer-aided design, automated assembly, computer numerically controlled machines, and group technology. The second major obstacle is the lack of customized software. Mr. Sgammato met with 15 different vendors and could not find software that completely matches his firm's needs. The basic problem is that the vendors simply do not understand his business. He laments the fact that he will probably have to hire expensive consultants to customize the software. Interestingly, he did not mention the cost of retraining workers in the aftermath of AMT implementation.

Mr. Sgammato was very excited about taking advantage of the State University of New York at Stony Brook's resources to help his firm overcome some of these obstacles. For example, in 1993, with financial support from the federal government, the Harriman School for Management and Policy at SUNY-Stony Brook initiated a "Jobs Project" or Dislocated Worker Training Program. The participants in this program were experienced engineers who had been laid off as a result of major cutbacks in defense programs on Long Island in the early 1990s. Most of these engineers had worked for the Grumman Corporation, by far the largest defense contractor on Long Island. The purpose of the Jobs Project was to retrain these individuals so they could be reemployed, presumably by firms in growing industries in the region, such as computers, environment and waste management, and biotechnology.

Under the supervision of Professor Gerrit Wolf and Mr. Joe Pufahl, the co-directors of the Jobs Project, a group of students agreed to assist Electronic Hardware with the selection and implementation of a computer-integrated manufacturing system. The group has visited the firm several times and has assisted Electronic Hardware in developing a budget for the CIM project. More importantly, the group has agreed to help the firm with some of the necessary customization of the software. This may obviate the need to hire consultants during the implementation phase. Mr. Sgammato was especially pleased about this. He also noted that his involvement with SUNY-Stony Brook had helped him become more aware of state-of-the-art manufacturing technologies that were developed at the engineering school. Finally, Mr. Sgammato also mentioned that his company had received assistance from professors and students at the SUNY-Farmingdale and Suffolk Community College.

SUMMARY OF FIELD INTERVIEWS

Several key findings emerged from my field research. First, there was strong evidence of "skill-upgrading," in the sense that firms invested heavily in retraining workers in the aftermath of technological innovation. Retraining is critical to these firms because computers are now ubiquitous on the factory floor, as the plant visits vividly demonstrated. Several managers showed us how the technologies had transformed the nature of work for many production workers. For example, workers must understand how to operate programmable machines using customized software; for many employees, this requires a new set of skills and responsibilities. The bottom line is that employees must possess a high level of skill in order to realize the full potential of AMTs. For example, if the firm uses automated technologies, its workers must select processing equipment, tooling, and job sequencing. Furthermore, they must be proficient in using the information that is generated by computers. The rate of return on investment in AMTs will depend, to some extent, on the quality of the workers who implement them. Therefore, it is critical for companies to also maximize the rate of return on their investment in human capital, through continuous training and "skill-building" programs and other efforts to raise the quality of worker output.

Our interviews of managers and workers suggest that certain elements of employee empowerment were enhanced in the aftermath of AMT adoptions. These include additional training, knowledge, and skill development, greater worker discretion and decision-making authority, and a move towards multi-skill and broader-scope jobs (referred to as job "enrichment" in the management literature [Daft 1998]). Consistent with the integrative aspect of many of the new technologies, AMT adoption also appears to be associated with the formation of teams. Some of these were cross-functional teams from engineering, production, logistics, R&D, procurement, and customer support.

Despite these somewhat dramatic changes for existing workers, it is important to note that AMTs have also led to downsizing because they reduce the demand for clerical and manual labor. Mechanization and a swift improvement in the flow of information has allowed these companies to significantly increase quality and productivity. Many of these AMTs are used to simplify and streamline administrative procedures.

An acceleration of the pace of technological change has dramatically changed the work environment. First, it has created anxiety about job security, particularly among middle managers. Technological change is certainly disruptive, but most of the downsizing has occurred through attrition. Second, workers now have more discretion (decision-making authority); however, this also entails more responsibility. The new technologies have also empowered workers, providing them with new skills and enabling them to learn more about other functional areas of the business.

. The cases also provided some interesting information on the obstacles to additional investment in AMTs. A pervasive finding is that companies find it difficult to quantify the benefits from technological investment. This is a major reason why firms have not implemented new technologies, despite the strong enthusiasm for them among the executives I interviewed. Almost all of these manufacturing experts agreed that the benefits outweigh the costs, but they cannot compute precise values. Furthermore, in instances where they generate rate-of-return projections, they rarely go back after the fact and assess how accurate those projections were. This often makes it difficult to justify further R&D projects. I suggest that they devote some additional effort to providing more precise figures on actual rates of return.

A second major obstacle/difficulty in AMT implementation is the quality of software. The largest expense in the implementation process was software, not hardware and physical equipment, because most companies were compelled to hire consultants to customize the software to fit the particular needs of the company. The greatest unfulfilled need of the firms in our sample was customized software.

The Electronic Hardware case provides a good example of how universities and government agencies can help firms overcome these barriers to implementation. The firm is utilizing an existing government/university program to improve its knowledge of state-of-the-art manufacturing technologies and develop more precise estimates of the costs and benefits of various AMTs. It is also drawing upon faculty and student expertise to help develop customized software for these new systems.

Despite the software difficulties experienced by many firms, I found strong evidence on our plant visits that AMTs have achieved their stated objectives. Recall that flexible manufacturing technologies involve the use of computers to coordinate machines and workers across functional areas of the business. Firms have been using computers on the factory floor for years. However, the separate implementation of computer-based technologies has not improved competitiveness as much as expected, because these stand-alone machines created new bottlenecks even as they removed old ones. Flexible manufacturing technologies were designed to eliminate these bottlenecks and streamline efficiency. It appears that they have done so, at least in the sample of firms we studied. As noted in the Pall Corporation case, AMTs have also allowed companies to enhance accounting and inventory controls.

Notes

1. See Siegel, Waldman, and Link (1999) for a more comprehensive discussion of qualitative research methods as they relate to field studies.
2. As noted in Kamer (1993), New York State lost 28% (65,000 jobs) of its defense employment base between 1985 and 1991.

7 Summary of Findings and Policy Implications

Two recent trends in the labor market have attracted a great deal of attention in academia and in the popular press. The first is the wave of large-scale corporate downsizing programs, coupled with a precipitous decline in the manufacturing workforce that began in the 1970s. A second trend is greater wage "inequality," or an increase in the wage premium associated with a college degree. This has occurred despite a large increase in the number of students who have been awarded college degrees in recent decades. A number of prominent labor economists have hypothesized that skill-biased technological change could be an important determinant of these trends.

Consequently, many authors have attempted to estimate the impact of technology on the demand for labor, usually attributing a great deal of quantitative significance to the role of skill-biased technological change in shifting the wage structure and labor composition. This evidence is remarkably robust to differences in methodology, selection of proxies for technological change, level of aggregation of the data, and choice of country. Yet, while these studies have been useful, they suffer from several important empirical limitations. Most authors have used industry-level data, indirect measures of technological change, and limited information on the composition of the labor force and changes in the work environment. This effectively precludes an in-depth analysis of determinants and consequences of technology adoption.

In this monograph, I have presented evidence from a detailed, firm-level survey that overcomes many of these limitations. The richness of the file enables me to provide comprehensive estimates of the impact of technological innovation on six classes of workers. I also examine the important question of whether the "skill-bias" of technological change varies for different classes of technologies; to my knowledge, this issue has not explored in any existing study.

SUMMARY OF STATISTICAL FINDINGS

The major conclusions of this study are as follows:

1. Technological change is associated with downsizing and a shift in labor composition in favor of workers with higher levels of education.

2. We can empirically distinguish between two classes of technologies: linked and integrated AMTs.

3. Downsizing and recomposition in favor of highly educated workers are more strongly associated with integrated AMTs. Thus, the nature of the skill bias may depend on the type of technology that is implemented.

4. Technology usage enhances employee empowerment. Empowerment appears to be more closely associated with linked, rather than integrated AMTs, again confirming the importance of conducting a disaggregated analysis of skill-bias technological change.

5. The two major obstacles to additional investment in new technologies are difficulties in quantifying benefits and the high cost of customizing software to fit company needs.

My first conclusion—that the implementation of a new technology is associated with corporate downsizing—is entirely plausible, because many of the new technologies were designed precisely to eliminate production and clerical jobs. Furthermore, these technologies promote integration, which implies that there will be an increase in the flow of information across functional areas within a company. This increase may also serve to eliminate jobs, since new technologies can help managers identify redundancies within the organization.

I also conclude that technology adoption is associated with skill upgrading. That is, it appears that these employment reductions are masking an important underlying shift in the demand for labor. Specifically, I find that technological change is associated with a shift in labor composition and compensation in favor of workers with higher levels of skill and education. These results are consistent with the idea that skill-upgrading occurs after new technologies are implemented on the factory floor. This finding is also not surprising, because the new technologies I examined typically involve the use of computers. Furthermore, because they are integrative, the new technologies often change the work environment; that is, in the aftermath of technological change, information and duties are more likely to be shared among workers performing different functional tasks. To be successful in this type of environment, workers must acquire or upgrade their computer skills and broaden their understanding of policies and procedures in other departments. The case studies and field interviews confirm these findings and describe how firms have invested in workers who can help them implement new technologies effectively.

These results are consistent with a study by Murnane, Willett, and Levy (1995), which attempts to explain the recent increase in the wage premium for educated workers. The authors note that (real) earnings of high school graduates have declined since 1979 in absolute terms and relative to the earnings of college graduates. They find that employers are requiring more workplace skills and are paying higher wages to those who possess them; also, employers are more selective in choosing employees. All of this has led to an increase in the returns on employees' skills. My results provide additional support for this argument, since it appears that the demand for highly educated labor increases after the implementation of a new technology.

I also find that technology is associated with certain elements of empowerment such as training, knowledge, skill development, and job enrichment. In some instances, technology also creates new job and career opportunities and provides employees with greater power or control and decision-making authority.

My findings also imply that there may be substantial heterogeneity across technologies in the impact on the demand for labor; specifically, the employment and recompositional effects differ by the type of technology, linked or integrated. A factor analysis confirms the validity of the distinction between linked and integrated AMTs, and I group the technologies accordingly. Not surprisingly, there is more downsizing and recompositioning of the labor force associated with integrated AMTs than with linked AMTs. On the other hand, linked AMTs appear to enhance employee empowerment more than integrated AMTs. Thus, when considering the employment and compositional effects of innovative activity, it may be important to consider the type of technological change.

Although this study focuses on the manufacturing sector, there is reason to suspect that the findings may have even stronger implications for service industries. As noted in Baily and Gordon (1988), service industries invest in computers at approximately double the rate of manufacturing industries. As in manufacturing, these computers are often used to streamline efficiency and raise quality. For example, "expert systems," or computer programs that simulate the behavior of human experts in business situations, are widely used in financial and communication services, especially in banking. Given that services are substantially more labor-intensive than manufacturing, the potential cost savings associated with the implementation of advanced technologies such as expert systems are quite high. Low-skilled service jobs appear to be particularly vulnerable, implying that the skill bias of technological change may be greater in services than in manufacturing.

POLICY IMPLICATIONS AND RECOMMENDATIONS

Technological change is indeed associated with skill upgrading and corporate downsizing. Qualitative evidence from our case reports and from previous studies reveals that new technologies require workers to develop new skills and to perform new tasks;[1] this is especially true for integrative technologies. The implementation of a new technology also appears to induce companies to invest additional resources in training and to enhance employee empowerment.

These changes in employment structure and in the work environment raise an important public policy issue relating to human capital development, because education is generally regarded as a public good and also an important input in the production of "skill." In fact, as noted in previous chapters, economists often use education as a proxy for skill. Therefore, we need to consider the appropriate role of government in addressing the consequences of technological change, which has spurred a large increase in the demand for skilled labor. An examination of the current imbalance between the supply and demand of skilled labor must begin with a consideration of the types of skills that employers value.

Recent modifications in skill requirements were well documented by Murnane and Levy (1996) in their monograph, *Teaching The New Basic Skills*. The authors reported that firms are increasingly demanding that workers possess what they term five "new basic skills." These are abilities to understand mathematical and analytical concepts, to solve semi-structured problems by formulating and testing hypotheses, to communicate effectively in oral and written forms, to work productively in groups with colleagues from diverse backgrounds, and to demonstrate proficiency with computers. The bottom line is that companies are insisting that workers have stronger cognitive and interactive skills.[2]

My results suggest that technical change further stimulates the demand for these skills, establishing an even greater wage premium for those who possess them. Common sense also dictates that integrative technologies require additional interactive skills, because information is exchanged among workers with different functional perspectives and expertise (e.g., manufacturing, R&D, accounting, finance, marketing). A critical issue that must be addressed by leaders in business, education, and government is whether our educational institutions are actually providing these skills.

Murnane and Levy (1996) suggested that primary and secondary schools are not teaching the new basic skills. This has caused overinvestment in higher education, because screening for such skills is expensive; so, many companies simply raise their educational requirements. The end result is that many firms are hiring college graduates for positions that normally require only a high school degree or vocational training, because they can be certain that college graduates actually possess these new basic skills.

Thus, firms are turning to the colleges (including community colleges) and universities for graduates whose skills complement new and existing technologies. In recent years, these institutions of higher learning have become more responsive to the needs of companies for several reasons. First, they have come to realize that firms are critical stakeholders. In fact, in some divisions of the university, especially the professional schools, administrators consider employers to be their primary "customers."[3] Second, it is important to note that universities receive substantial direct and indirect funds and subsidies from government.[4] It is very useful to have the business community as an ally when lobbying the state legislature for additional financial support.

Also, federal agencies, such as the National Science Foundation, have strongly encouraged university/industry interactions in their funding policies. It is now quite common for researchers from academia and industry to jointly submit research proposals to NSF. This is part of an overall initiative to enhance domestic competitiveness by placing a greater emphasis on applied research. NSF has also supported manufacturing centers at various locations so that companies can utilize the most advanced manufacturing technologies.[5] The end result is that universities (the "suppliers") are helping companies (the "customers") enhance the skills of their employees.[6]

This emphasis on university-industry cooperation stems primarily from two important laws: the Bayh-Dole University and Small Business Patent Act (1980) and the Omnibus Trade and Competitiveness Act (1988). The 1980 legislation facilitated interaction between universities and private firms by allowing grantees and contractors receiving federal R&D funds (such as small businesses, universities, and other non-for-profit institutions) to license new technologies. The 1988 legislation created several new programs in the Department of Commerce's National Institute of Standards and Technology (NIST), such as Manufacturing Technology Centers, to help small and medium-size manufacturers become more competitive.

Despite this assistance, firms must also confront skill "shortages" for existing and prospective employees through on-the-job training, which is expensive and time-consuming. The traditional paradigm is that companies hire technical graduates and subsequently provide them with on-the-job-training to develop their competence with new equipment and production processes. Given the rapid pace of technological change and an increase in the intensity of global competition, many firms are discovering that their demand for trained technicians is growing faster than their ability to furnish the appropriate hands-on training time.[7] These shortages are readily apparent in high tech regions such as Silicon Valley and are spreading across the country.

One notable development in alleviating the skills shortage has been the establishment of new and creative forms of alliances between universities and private firms. For example, Arizona State University has recently launched a joint venture with Intel and Motorola to construct a microelectronics teaching factory. This facility will serve as a simulated version of a real microchip factory. Industry and university officials project that this facility will help solve the industry's need for skilled technical and management teams. By partnering with the university, Intel and Motorola can further the technical education of permanent members of the work force and prospective employees.

It is important to note that substantial spillovers could result from the construction of these types of facilities. For example, customers and suppliers of Intel and Motorola, who number in the thousands in the local Phoenix area alone, will also benefit from this initiative. The diffusion of these new technologies could stimulate the local economy, through the creation of start-up companies, new jobs, and additional investment in R&D.

Another pragmatic way to help alleviate the skills shortage is to design policies that help promote an increase in the <u>supply</u> of skills, especially those that complement the implementation of new technologies. One recommendation is for state and local government officials to provide more incentives for the development of alliances between colleges and universities and firms. These could involve the use of tax incentives and additional assistance in helping companies surmount some of the financial and regulatory barriers that often serve as obstacles in strengthening these relationships.

A second recommendation is to strengthen the connection between classroom instruction and the skill requirements of employers. As a first step, administrators at institutions of higher learning must pay closer attention to the needs of employers. The implications of the findings presented in Murnane and Levy is that there is a "market" failure in the delivery of critical skills. Existing evidence suggests a growing demand for cognitive and interactive skills, which have not been adequately supplied by the educational establishment.

The most effective way for colleges and universities to monitor the provision of skills is for these institutions to develop assessment centers. Assessment centers are units that are expressly designed to quantitatively measure and evaluate the skills and abilities or workers and job candidates. These centers are usually staffed by applied (industrial/organizational) psychologists and human resource management professionals.

As discussed by Riggio and Mayes (1997), assessment centers were first employed in the private sector to screen and evaluate candidates for top management positions. Later, they were used as a career development and training tool at all levels of the corporation. In a managerial context, workers are typically evaluated along such dimensions as leadership, interpersonal skills, organization and planning, decisiveness, decision making, perception, and oral and written communication.[8]

In recent years, assessment centers have been used in educational settings to evaluate undergraduate students in business administration (see Riggio et al. 1997), graduate students in applied psychology (Kottke and Shultz 1997), and students in other majors (Rea, Rea, and Moomaw 1990; Wendel and Joekel 1991). Many colleges and universities have been moving in this direction by conducting outcomes assessments for individual courses, which involves measuring the "value-added" of an individual instructor.

I propose that the use of assessment centers be expanded greatly at all levels of education. A first step in improving the "productivity" of education is effective measurement and evaluation of educational outcomes. Assessment centers will help focus greater attention on the "production process," or in this case, the delivery and absorption of knowledge and skill. A quantitative approach is also quite useful because it facilitates benchmarking across schools and over time. Also, it will help ensure accountability for the quality of output (graduates) and permit firms to provide useful feedback to educators.

Finally, there are two reasons why it is especially important that a major effort be made to provide for the needs of <u>small</u> companies. First, skill shortages appear to be most acute for these firms.[9] Many of the smaller, non-defense related firms that I visited cited a shortage of skilled workers as an impediment to further investment in technology. Second, small firms create a substantial number of net new jobs in many industries. Programs that reduce these barriers to growth would be especially helpful in promoting economic development. Community colleges also play an important role in high-technology workforce development and may be particularly useful in targeting smaller companies.

In this section, I have outlined some of the policy implications of the empirical findings. My two recommendations are that educational institutions develop closer alliances with businesses and nonprofit organizations that employ their graduates and that they use an assessment center approach to evaluate educational outcomes.

The first recommendation is consistent with what I consider to be a healthy trend towards closer customer-supplier relationships in the private sector; that is, firms are "consuming" the output of educational institutions and thus can provide much-needed feedback to improve educational performance. The second recommendation underscores the importance of the measurement and evaluation of educational services, especially in an economy that is increasingly knowledge-based. I do not believe there is much mystery here. As noted by Murnane and Levy (1996), employers, academics, and consultants have explicitly identified these critical skills. It is incumbent upon our educational institutions to ensure that these skills are delivered.

It is also important to note that my recommendations could enable firms to surmount barriers to additional technological investment. Recall that the firms we interviewed had two major concerns: difficulties in quantifying benefits of the new computerized technologies and problems with customizing software. The field research revealed that university/government/industry partnerships, such as the federally funded Jobs Project at SUNY-Stony Brook, can help firms deal with these concerns by providing valuable consulting advice that can help them justify, design, and implement new technologies.

The bottom line is that the proposed measures would spur additional investment in new technology. They would also reduce the need for firms to invest in on-the-job training and other expensive means of skills enhancement. The evidence is clear that employers value workers who can help them implement new technologies. Society gains when there is a more rapid rate of technological diffusion. Thus, to ensure a continual rise in our standard of living, educational institutions must make appropriate adjustments to ensure the delivery of a workforce whose skills complement technological change.

Notes

1. See Adler (1986, 1988), Dean and Snell (1991), and Snell and Dean (1992).
2. Most economists do not actually measure "skill" (mea maxima culpa!); instead, they often use education as a proxy for skill. For studies that present actual mea-

sures of changes in skill requirements, see Howell and Wolff (1991, 1992) and Murnane, Willett, and Levy (1995).

3. For example, many large, multinational corporations have played a leading role in assisting America's leading graduate schools of business in their efforts to modify their curricula by emphasizing the importance of global perspectives and proficiency with computers.

4. It is important to consider the appropriate role of the government. On the one hand, policymakers wish to eliminate obstacles to additional investment in R&D. On the other hand, it is socially inefficient for the government to subsidize efforts that firms would fully fund under normal conditions.

5. The establishment of these centers grew out of a concern that American firms would not be competitive with foreign companies unless they were using the most advanced manufacturing technologies.

6. These partnerships can be viewed as a joint venture between customer and supplier, as opposed to a typical joint venture between two competitors or firms in related industries.

7. This is due, in part, to a shortage of experienced trainers. Some firms have found that training costs are prohibitive, so they have avoided the use of new technologies.

8. See Riggio and Mayes (1997) for further details.

9. Many small companies also have a legitimate fear of dealing with large, bureaucratic institutions (such as the university), since they often do not have the appropriate infrastructure to contend with the necessary paperwork. Any assistance in this regard would also be greatly appreciated.

Appendix

Survey Questionnaire

CIMS-Computer Integrated Manufacturing Systems

1. **Please classify the extent of usage of CIMS technologies (defined on the reverse side of the cover letter) in your Long Island manufacturing operations by checking the appropriate row.**

	AAS	AS/R	AGVs	CAD	CAM	CNC	FMS	GT	JIT	MRP	ROB	SPC
(1) Never formally considered												
(2) Considered but not adopted												
(3) Under active consideration or soon to be implemented												
(4) Implemented and in use												
(5) Implemented but terminated												

If you are unsure which column to use for some CIMS technologies that you have considered or implemented, please specify the technologies and the columns to which you assigned them.

If all your responses are in rows 1 and 2 skip to question 8. Otherwise continue to question 2.

2. **For each technology you have implemented (Rows 4 and 5 above), please provide the following information.**

	AAS	AS/R	AGVs	CAD	CAM	CNC	FMS	GT	JIT	MRP	ROB	SPC
Year of decision to adopt												
First year in use												
Cost of investment (including training)												
Investment made with internal funds (yes/no)												
Investment made with other funds (specify source)												
Year of termination (if applicable)												

3. For each technology you have implemented check the <u>primary reasons for adoption</u>. Multiple reasons can be checked.

	AAS	AS/R	AGVs	CAD	CAM	CNC	FMS	GT	JIT	MRP	ROB	SPC
Increased sales growth												
Adoption of CIMS by competitors												
Cost reduction												
Increased responsiveness to customer needs												
Quality improvement												
Increased flexibility (design or manufacturing)												
Adoption of CIMS by customers or suppliers												
Federal funding of technology												
Increased managerial control												
Increased employee control												
Other (please specify)												

4. For each technology you have implemented, check the <u>significant results</u>. Multiple results can be checked.

	AAS	AS/R	AGVs	CAD	CAM	CNC	FMS	GT	JIT	MRP	ROB	SPC
Increased sales growth												
Adoption of CIMS by competitors												
Cost reduction												
Increased responsiveness to customer needs												
Quality improvement												
Increased flexibility (design or manufacturing)												
Adoption of CIMS by customers or suppliers												
Federal funding of technology												
Increased managerial control												
Increased employee control												
Other (please specify)												

5. For each technology you have implemented, check the **methods of implementation**. Multiple methods can be checked.

	AAS	AS/R	AGVs	CAD	CAM	CNC	FMS	GT	JIT	MRP	ROB	SPC
Use of an outside vendor												
Training existing personnel												
Hiring new personnel												
Laying off personnel (please provide % in parentheses)												
Reorganizing reporting relationships												
Changing employees' job responsibilities												
Increased control and monitoring by management												
Create new jobs and career opportunities for employees												
Other (please specify)												

6. For each technology you have implemented, check the **functions significantly affected** by adoption. Multiple functions can be checked.

	AAS	AS/R	AGVs	CAD	CAM	CNC	FMS	GT	JIT	MRP	ROB	SPC
Marketing/sales												
Accounting/finance												
R&D												
Product design/engineering												
Production (inc. planning, inventory)												
Other (please specify)												

7. For each technology you have implemented and terminated, check the <u>factors responsible for termination</u>. Multiple factors can be checked.

	AAS	AS/R	AGVs	CAD	CAM	CNC	FMS	GT	JIT	MRP	ROB	SPC
Insufficient employee training												
Personnel problems												
Union problems												
Lack of managerial cooperation												
High operating costs												
System failed or was unreliable												
Other (please specify)												

8. For <u>each technology you considered in detail but did not implement</u>, please check the reasons for your decision. Multiple reasons can be checked.

	AAS	AS/R	AGVs	CAD	CAM	CNC	FMS	GT	JIT	MRP	ROB	SPC
Insufficient internally generated funds												
Difficult to raise funds												
Difficult to quantify benefits												
Upfront costs of hardware & training												
Problems with cost accounting methods												
Union problems												
Other (please specify)												

9. What percentage of your company's manufacturing employees (not including outsourcing) work:

On Long Island? _____%
Outside N.Y.S. but in the U.S.? _____%

Off Long Island but in New York State? _____%
Outside U.S.? _____%

10. Are there any CIMS technologies that you use off Long Island (L.I.) but not on L.I.?

No _____
Yes _____

If yes, which technologies, and why weren't the technologies implemented on L.I.?

_____

_____

_____

BACKGROUND QUESTIONS:

11. List in descending order of sales the 3-digit SIC codes that best describe your products. List the SIC code associated with your primary product first (see list of SIC codes on reverse side of cover letter)

1. _____ 2. _____ 3. _____ 4. _____ 5. _____

12. What percentage of your Long Island production, as measured by sales dollars, is shipped to:

a.	Other firms on Long Island	_____%	e.	Firms in the U.S. outside New York State	_____%
b.	Other firms in New York State but off L.I.	_____%	f.	Customers outside the U.S.	_____%
c.	Other divisions in your firm on L.I.	_____%	g.	Defense industry customers	_____%
d.	Other divisions in your firm off L.I.	_____%			

13. To approximately how many L.I. customers do you ship? _____

14. What was the 1989 gross dollar value of sales from your Long Island operations:

$_____

15. What percentage of your Long Island labor force turned over during 1989? _____%

16. In what year did your company begin operations on Long Island? _____

17. Has your company been acquired by another firm within the last three years? _____

18. What percentage of your purchases is from:

a.	L.I. operations of other firms	_____%	e.	U.S. firms outside New York State	_____%
b.	Firms in New York State but off L.I.	_____%	f.	Sources outside U.S.	_____%
c.	Other L.I. divisions of your firm	_____%	g.	Defense industry firms	_____%
d.	Other divisions of your firm off L.I.	_____%			

19. From approximately how many L.I. firms do you purchase? _____

20. How many people are employed in your Long Island operations?

	Currently	12/89	12/88	12/87
Total				
Managerial & Supervisory				
Technical & Professional				
Clerical & Administrative				
Direct Labor & Supporting Personnel				
Other				

21. What percentage of your company's revenue is spent on R&D? _____%

22. How many persons in your Long Island operations were performing R&D as significant parts of their job as of:

	Currently	12/89	12/88	12/87
Scientists and Engineers				
Others				

23. What percentage of the R&D budget for your company is financed by the federal government? _____%

24. What percentage of the R&D performed by your company is:

basic research	_____%	process innovation	_____%
applied research	_____%	other	_____%
product innovation	_____%		

25. What percentage of your R&D is performed on Long Island?   _____%

26. There may be CIMS implementation or effectiveness issues that were overlooked in this questionnaire or on which you would like to comment. If so, please use additional pages.

YOUR NAME, TITLE AND PHONE NUMBER

_____
(Name)

_____
(Title)

_____
(Phone Number)

References

Adler, Paul S. 1986. "New Technologies, New Skills." *California Management Review* 29(1): 9–28.

Adler, Paul. S. 1988. "Managing Flexible Automation." *California Management Review* 30(3): 34–56.

Appelbaum, Eileen, and Rosemary Batt. 1994. *The New American Workplace: Transforming Work Systems in the United States.* Ithaca, New York: ILR Press.

Autor, David H., Lawrence F. Katz, and Alan B. Krueger. 1997. *Computing Inequality: Have Computers Changed The Labor Market?* NBER working paper no. 5956, National Bureau of Economic Research, Cambridge, Massachusetts.

Ayres, Ralph. 1988. "Future Trends in Factory Automation." *Manufacturing Review* 1: 93–103.

Baily, Martin N., and Robert J. Gordon. 1988. "The Productivity Slowdown, Measurement Issues, and the Explosion of Computer Power." *Brookings Papers on Economic Activity* (2): 347–420.

Bartel, Ann P. 1994. "Productivity Gains from the Implementation of Employee Training Programs." *Industrial Relations* 33(4): 411–425.

Bartel, Ann P. 1995. "Training, Wage Growth, and Job Performance: Evidence from a Company Database." *Journal of Labor Economics* 13(3): 401–425.

Bartel, Ann P., and Frank R. Lichtenberg. 1987. "The Comparative Advantage of Educated Workers in Implementing New Technology." *Review of Economics and Statistics* 69: 1–11.

Bartel, Ann P., and Frank R. Lichtenberg. 1990. "The Impact of Age of Technology on Employee Wages." *Economics of Innovation and New Technology* 1: 1–17.

Bartel, Ann P., and Nachum Sicherman. 1999. "Technological Change and Wages: An Interindustry Analysis." *Journal of Political Economy* 107(2): 285–325.

Batt, Rosemary, and Eileen Appelbaum. 1995. "Worker Participation in Diverse Settings: Does the Form Affect the Outcome, and If So, Who Benefits?" *British Journal of Industrial Relations* 33(3): 353–378.

Berman, Eli, John Bound, and Zvi Griliches. 1994. "Changes in the Demand for Skilled Labor within U.S. Manufacturing Industries: Evidence from the Annual Survey of Manufacturing." *Quarterly Journal of Economics* 109: 367–397.

Berman, Eli, John Bound, and Stephen Machin. 1998. "Implications of Skill-Biased Technological Change: International Evidence." *Quarterly Journal of Economics* 112(4): 1245–1279.

Berndt, Ernst R., Catherine J. Morrison, and Larry S. Rosenblum. 1992. *High Tech Capital Formation and Labor Composition in U.S. Manufacturing Industries: An Exploratory Analysis.* NBER working paper no. 4010, National Bureau of Economic Research, Cambridge, Massachusetts.

Betts, Julian R. 1997. "The Skill Bias of Technological Change in Canadian Manufacturing Industries." *Review of Economics and Statistics* 79(1): 146–150.

Black, Sandra E., and Lisa Lynch. 1997. *How to Compete: The Impact of Workplace Practices and Information Technology on Productivity.* NBER working paper no. 6120, National Bureau of Economic Research, Cambridge, Massachusetts.

Black, Sandra E., and Lisa Lynch. 1996. "Human Capital Investments and Productivity." *American Economic Review* 86(2): 263–267.

Bound, John, and George Johnson. 1992. "Changes in the Structure of Wages during the 1980's: An Evaluation of Alternative Explanations." *American Economic Review* 82: 371–392.

Bound, John, David A. Jaeger, and Regina M. Baker. 1995. "Problems with Instrumental Variables Estimation When the Correlation between the Instruments and the Endogenous Explanatory Variable is Weak." *Journal of the American Statistical Association* 90(430): 443–450.

Bresnahan, Timothy F., and Manuel Trajtenberg. 1995. "General Purpose Technologies." *Journal of Econometrics* 65: 83–108.

Brynjolfsson, Erik, Thomas Malone, Vijay Gurbaxani, and Ajit Kambil. 1994. "Does Information Technology Lead to Smaller Firms?" *Management Science* 40: 1628–1644.

Carlsson, Bo, David B. Audretsch, and Zoltan Acs. 1994. "Flexible Technology and Plant Size: U.S. Manufacturing and Metalworking Industries." *International Journal of Industrial Organization* 13(12): 359–372.

Chase, Richard B., and N.J. Aquilano. 1995. *Production and Operations Management: Manufacturing and Services.* Chicago: Richard D. Irwin.

Chen, F. Frank, and Everett E. Adam, Jr. 1991. "The Impact of Flexible Manufacturing Systems on Productivity and Quality." *IEEE Transactions on Engineering Management* 38(1): 33–45.

Chennells, Lucy, and John Van Reenen. 1995. "Wages and Technology in British Plants: Do Workers Get a Fair Share of the Plunder?" Paper presented at the National Academy of Sciences Conference on Science, Technology, and Economic Growth, Washington, D.C., May.

Cohen, Morris A., and Uday M. Apte. 1997. *Manufacturing Automation.* Chicago: Irwin.

Cohen, Wesley M., and Daniel Levinthal. 1989. "Innovation and Learning: The Two Faces of R&D." *The Economic Journal* 99: 569–596.

Conger, Jay Allen, and R.N. Kanungo. 1988. "The Empowerment Process: Integrating Theory and Practice." *Academy of Management Review* 13: 471–482.

Daft, Richard L. 1998. *Organizational Theory and Design,* 6th ed. Cincinnati, Ohio: Southwestern College Publishing.

Davis, Steven J., and John Haltiwanger. 1991. "Wage Dispersion between and within U.S. Manufacturing Plants, 1963–1986." *Brookings Papers on Economic Activity: Microeconomics,* pp. 115–200.

Davenport, Thomas. H. 1993. *Process Innovation: Reengineering Work through Information Technology.* Boston: Harvard Business School Press.

De Pietro, Rocco A., and Gina Massaro Schremser. 1987. "The Introduction of Advanced Manufacturing Technology (AMT) and Its Impact on Skilled Workers' Perceptions of Communication, Interaction, and Other Job Outcomes at a Large Manufacturing Plant." *IEEE Transactions on Engineering Management* 34(1): 4–11.

Dean, Edwin, Kent Kunze, and Larry Rosenblum. 1988. "Productivity Change and the Measurement of Heterogeneous Labor Inputs." Paper presented at the U.S.-Israeli Department-to-Ministry Seminar Program, Washington, D.C., October.

Dean, James W. 1987. *Deciding to Innovate.* Cambridge, Massachusetts: Ballinger Publishing.

Dean, James W., and Scott A. Snell. 1991. "Integrated Manufacturing and Job Design: Moderating Effects of Organizational Inertia." *Academy of Management Journal* 34(4): 776–804.

Dean, James W., and Scott A. Snell. 1996. "The Strategic Use of Integrated Manufacturing: An Empirical Examination." *Strategic Management Journal* 17(6): 459–480.

Delaney, John, Casey Ichniowski, and David Lewin. 1989. *Human Resource Management Policies and Practices in American Firms.* Bureau of Labor Management Relations and Cooperative Programs report, BLMR no. 137. Washington, D.C.: Government Printing Office.

Denison, Edward F. 1962. *The Sources of Economic Growth in the U.S. and the Alternatives before Us.* Supplementary paper no. 13, New York: Committee for Economic Development.

DiNardo, John E., and Jorn-Steffen Pischke. 1997. "The Returns to Computer Use Revisited: Have Pencils Have Changed the Wage Structure Too?" *Quarterly Journal of Economics* 112: 291–303.

Doms, Mark, Timothy Dunne, and Mark J. Roberts. 1994. "The Role of Technology Use in the Survival and Failure of Manufacturing Plants." Photocopy. Center for Economic Studies, U.S. Bureau of the Census.

Doms, Mark, Timothy Dunne, and Kenneth R. Troske. 1997. "Workers, Wages, and Technology." *Quarterly Journal of Economics* 112: 253–290.

Dunne, Timothy. 1994. "Plant Age and Technology Usage in U.S. Manufacturing Industries." *Rand Journal of Economics* 25(3): 488–499.

Dunne, Timothy, John Haltiwanger, and Kenneth R. Troske. 1996. *Technology and Jobs: Secular Change and Cyclical Dynamics.* NBER working paper no. 5656, National Bureau of Economic Research, Cambridge, Masssachusetts.

Dunne, Timothy, Mark J. Roberts, and Larry Samuelson. 1989. "The Growth and Failure of U.S. Manufacturing Plants." *Quarterly Journal of Economics* 104: 671–698.

Dunne, Timothy, and James A. Schmitz. 1995. "Wages, Employer Size-Wage Premia and Employment Structure: Their Relationship to Advanced Technology Usage in U.S. Manufacturing Establishments." *Economica* 62: 89–107.

Entorf, Horst, and Francis Kramarz. 1995. "The Impact of New Technologies on Wages and Skills: Lessons from Matching Data on Employees and on Their Firms." Paper presented at the National Academy of Sciences Conference on Science, Technology, and Economic Growth, May.

Ettlie, John E., and Ernesto Reza. 1992. "Organization Integration and Process Innovation." *Academy of Management Journal* 35: 795–827.

Fine, Charles H. 1993. "Developments in Manufacturing Technology and Economic Evaluation Models." In *Logistics of Production and Inventory, North-Holland Series of Handbooks in Operations Research and Management Science*, S.C. Graves, A.H.G. Rinnooy Kan, and P.H. Zipkin eds. Amsterdam: North Holland-Elsevier Science, pp. 711–750.

Foston, Arthur, Carolena L. Smith, and Tony Au. 1991. *Fundamentals of Computer Integrated Manufacturing.* Engelwood Cliffs: Prentice Hall.

Goldin, Claudia, and Lawrence F. Katz. 1996. "Technology, Skill, and the Wage Structure: Insights from the Past." *American Economic Review* 86(2): 252–257.

Gorsuch, Richard L. 1974. *Factor Analysis.* Philadelphia: W.B. Saunders.

Gould, Stephen Jay. 1981. *The Mismeasure of Man.* New York: W.W. Norton & Co.

Griliches, Zvi. 1969. "Capital-Skill Complementarity." *Review of Economics and Statistics* 51(4): 465–468.

Griliches, Zvi. 1970. "Notes on the Role of Education in Production Functions and Growth Accounting." In *Education and Income*, Vol. 35, NBER

Studies in Income and Wealth, Lee Hansen, ed. New York: Columbia University Press.

Griliches, Zvi. 1996. *Education, Human Capital, and Growth: A Personal Perspective.* NBER working paper no. 5426, National Bureau of Economic Research, Cambridge, Massachusetts.

Griliches, Zvi. 1998. *R&D and Productivity: The Econometric Evidence.* National Bureau of Economic Research for the University of Chicago Press. Chicago: University of Chicago Press.

Groshen, Erica L. 1990. *Employers, Occupations, and Wage Inequality in Three Cities 1957–1989: Another Piece of the Puzzle.* Working paper, Federal Reserve Bank of Cleveland, October.

Hannan, Timothy, and John M. McDowell. 1984. "The Determinants of Technology Adoption: The Case of the Banking Firm." *Rand Journal of Economics* 15: 328–353.

Haskel, Jonathan. 1999. "Small Firms, Contracting-Out, Computers, and Wage Inequality: Evidence from U.K. Manufacturing." *Economica* 66(1): 1–21.

Haskel, Jonathan, and Yiva Heden. 1999. "Computers and the Demand for Skilled Labour: Industry and Establishment-Level Panel Evidence for the U.K." *Economic Journal* 109 (March): 68–79.

Heizer, Jay, and Barry Render. 1999. *Operations Management 5th ed.* Upper Saddle River, New Jersey: Prentice-Hall.

Helper, Susan. 1999. "Complementarity and Cost Reduction: Evidence from the Auto Supply Industry." Paper presented at the National Bureau of Economic Research Conference on Organizational Change and Performance Improvement, Santa Rosa, California.

Hildreth, Andrew K.G. 1998. "Rent-Sharing and Wages: Product-Demand or Technology Driven Premia?" *Economics of Innovation and New Technology* 5(2): 199–226.

Hollen, C.R., and G.N. Rogol, 1985. "How Robotization Affects People." *Business Horizons* (May-June): 74–80.

Howell, David R., and Edward N. Wolff. 1991. "Trends in the Growth and Distribution of Skills in the U.S. Workplace, 1960–1985." *Industrial and Labor Relations Review* 44(3): 486–502.

Howell, David R., and Edward N. Wolff. 1992. "Technical Change and the Demand for Skills by U.S. Industries." *Cambridge Journal of Economics* 16: 127–146.

Ichniowski, Casey, Kathryn Shaw, and Giovanna Prennushi. 1997. "The Effects of Human Resource Management Practices on Productivity: A Study of Steel Finishing Lines." *American Economic Review* 87(3): 291–313.

Jensen, Richard. 1988. "Information Cost and Innovation Adoption Policies." *Management Science* 34(February): 230–239.

Jorgenson, Dale W., Frank W. Gollop, and Barbara Fraumeni. 1987. *Productivity and U.S. Economic Growth, 1979–1985*. Cambridge: Harvard University Press.

Jorgenson, Dale W., and Zvi Griliches. 1967. "The Explanation of Productivity Change." *Review of Economic Studies* 34: 249–283.

Jovanovic, Boyan. 1982. "Selection and Evolution of Industry." *Econometrica* 50: 649–670.

Juhn, Chinhui, Kevin M. Murphy, and Brooks Pierce. 1993. "Wage Inequality and the Rise in Returns to Skill." *Journal of Political Economy* 101(3): 410–442.

Juran, Joseph M. 1989. *Juran on Leadership for Quality: An Executive Handbook*. Wilson, Connecticut: Juran Institute.

Kamer, Pearl M. 1993. "Long Island's Defense Sector: Surviving the Current Defense Build-Down." Photocopy. Long Island Association, Commack, N.Y.

Katz, Lawrence F., and Kevin Murphy. 1992. "Changes in the Relative Wages, 1963–1987: Supply and Demand Factors." *Quarterly Journal of Economics* 107: 35–78.

Kelley, MaryEllen R., and Harvey Brooks. 1991. "External Learning Opportunities and the Diffusion of Process Innovations to Small Firms." *Technological Forecasting and Social Change* 39: 103–125.

Kelley, MaryEllen R. 1994. "Productivity and Information Technology." *Management Science* 40(11): 1406–1425.

Klein, Janice A. 1991. "A Reexamination of Autonomy in Light of New Manufacturing Practices." *Human Relations* 44: 21–38.

Kottke, Janet L., and Kenneth S. Shultz. 1997. "Using an Assessment Center as a Developmental Tool for Graduate Students: A Demonstration." *Journal of Social Behavior and Personality* 12(5): 289–302.

Krueger, Alan B. 1993. "How Computers Have Changed the Wage Structure: Evidence from Microdata." *Quarterly Journal of Economics* 108: 33–61.

Kusiak, Andrew. 1985. "Flexible Manufacturing Systems: A Structural Approach." *International Journal of Production Research* 23(6): 1057–73.

Lawler, Edward E., III, Susan Mohrman, and Gerald Ledford. 1992. *Employee Involvement and TQM: Practice and Results in Fortune 5000 Companies*. San Francisco: Jossey-Bass.

Lawrence, Robert Z., and Matthew J. Slaughter. 1993. "International Trade and American Wages in the 1980's: Giant Sucking Sound or Small Hiccup?" *Brookings Papers on Economic Activity, Microeconomics*, pp. 161–210.

Lee, Lei-Fung., G.S. Maddala, and Robert P. Trost. 1979. "Asymptotic Covariance Matrices of Two-Stage Probit and Two-Stage Tobit Methods for Simultaneous Equations Models with Selectivity." *Econometrica* 48: 491–503.

Leonard-Barton, Dorothy. 1992. "The Factory as a Learning Laboratory." *Sloan Management Review* 34(1): 23–38.

Levin, Sharon, Stanford L. Levin, and John B. Meisel. 1987. "A Dynamic Analysis of the Adoption of a New Technology: The Case of Optical Scanners." *Review of Economics and Statistics* 69: 12–17.

Levy, Frank, and Richard J. Murnane. 1992. "U.S. Earnings Levels and Earning Inequality: A Review of Recent Trends and Proposed Explanations." *Journal of Economic Literature* 30(3): 1333–1381.

Lichtenberg, Frank R., and Donald Siegel. 1991. "The Impact of R&D Investment on Productivity—New Evidence Using Linked-R&D-LRD Data." *Economic Inquiry* 29(3): 203–229.

Link, Albert N. 1987. *Technological Change and Productivity Growth.* Fundamentals of Pure and Applied Economics, No. 13. Chur, Switzerland: Harwood Academic Publishers.

Lynch, Lisa M., and Paul Osterman. 1989. "Technological Innovation and Employment in Telecommunications." *Industrial Relations* 28(2): 188–205.

MacDuffie, John Paul, Kannan Sethuraman, and Marshall L. Fisher. 1996. "Product Variety and Manufacturing Performance: Evidence from the International Automotive Assembly Plant Study." *Management Science* 42(3): 350–369.

Machin, Stephen. 1996. "Changes in the Relative Demand for Skills in the U.K. Labour Market." In *Acquiring Skills*, Allison Booth and Dennis Snower, eds. Cambridge: Cambridge University Press.

Maddala, G.S. 1983. *Limited-Dependent and Qualitative Variables in Econometrics.* Cambridge: Cambridge University Press.

Malthus, Thomas R. 1978. *Population: The First Essay.* Ann Arbor, Michigan: University of Michigan Press.

Markland, Robert E., Shawnee K. Vickrey, and Robert A. Davis. 1998. *Operations Management: Concepts in Manufacturing and Services,* 2nd ed. Cincinnati, Ohio: Southwestern College Publishing.

Marx, Karl. 1967. *Das Kapital.* New York: International Publishers, Co. Inc.

McCardle, Kevin. 1985. "Information Acquisition and the Adoption of New Technology." *Management Science* 31: 1372–1389.

McGuckin, Robert H., Mary Streitweiser, and Mark Doms. 1998. "The Effect of Technology Use on Productivity Growth." *Economics of Innovation and New Technology* 7(1): 1–26.

Mincer, Jacob. 1989. *Human Capital Responses to Technological Change in the Labor Market*. NBER working paper no. 3581, National Bureau of Economic Research, Cambridge, Massachusetts.

Mishel, Lawrence, and Jared Bernstein. 1994. "Is the Technology Black Box Empty? An Empirical Examination of the Impact of Technology on Wage Inequality and the Employment Structure." Photocopy, Economic Policy Institute: Washington, D.C.

Morrison, Catherine J., and Donald Siegel. 1996. "The Impact of Technology, Trade, and Outsourcing on Employment and Labor Composition." Photocopy. July.

Murnane, Richard J., and Frank Levy. 1996. *Teaching the New Basic Skills*. New York: The Free Press.

Murnane, Richard J., John B. Willett, and Frank Levy. 1995. "The Growing Importance of Cognitive Skills in Wage Determination." *Review of Economics and Statistics* 77: 251–266.

Murphy, Kevin M., and Finis Welch. 1992. "The Structure of Wages." *Quarterly Journal of Economics* 107: 215–226.

National Science Foundation. 1989. *Research and Development in Industry*, serial 1957–1989 (data series), Washington, D.C.: Government Printing Office.

Nelson, Richard R., and Edmund S. Phelps. 1966. "Investment in Humans, Technological Diffusion, and Economic Growth." *American Economic Review* 56: 69–75.

New York Stock Exchange Office of Economic Research. 1982. *People and Productivity: A Challenge to Corporate America*. New York: New York Stock Exchange, Office of Economic Research.

Osterman, Paul. 1986. "The Impact of Computers on the Employment of Clerks and Managers." *Industrial and Labor Relations Review* 39(2): 175–186.

Pakes, Ariel, and Richard Ericson. 1989. *Empirical Implications of Alternative Models of Firm Dynamics*. NBER working paper no. 2893, National Bureau of Economic Research, Cambridge, Massachusetts.

Park, Ki Seong. 1996. "Economic Growth and Multiskilled Workers in Manufacturing." *Journal of Labor Economics* 12(2): 254–285.

Rea, P., J. Rea, and C. Moomaw. 1990. "Skills Development." *Personnel Journal* 69(4): 126–131.

Regev, Haim. 1995. "Innovation, Skilled Labor, Technology and Performance among Israeli Industrial Firms." Paper presented at the National Academy of Sciences Conference on Science, Technology, and Economic Growth, May.

Reilly, Kevin T. 1995. "Human Capital and Information." *Journal of Human Resources* 30: 1–18.

Revenga, Ana. 1992. "Exporting Jobs" The Impact of Import Competition on Employment and Wages in U.S. Manufacturing." *Quarterly Journal of Economics* 107(1): 255–284.

Ricardo, David. 1995. *Principles of Political Economy.* Chicago: Richard D. Irwin.

Riggio, Ronald E., Monica Aguirre, Bronston T. Mayes, Chris Belloli, and Carolyn Kubiak. 1997. "The Use of Assessment Center Methods for Student Outcome Assessment." *Journal of Social Behavior and Personality* 12(5): 273–288.

Riggio, Ronald E., and Bronston T. Mayes, eds. 1997. Assessment Centers: Research and Application [Special Issue]. *Journal of Social Behavior and Personality* 12(5): 273–288.

Robot Institute of America. 1986. *Worldwide Survey and Directory on Industrial Robots.* Robot Institute of America: Dearborn, Michigan.

Romeo, Anthony. 1975. "Interindustry and Interfirm Differences in the Rate of Diffusion of an Innovation." *Review of Economics and Statistics* 57: 311–319.

Schlie, Theodore, and Joel Goldhar. 1989. "Product Variety and Time Based Manufacturing and Business Management: Achieving Competitive Advantage through CIM." *Manufacturing Review* 2: 32–42.

Schonberger, Richard J. 1982. *Japanese Manufacturing Techniques: Nine Hidden Lessons in Simplicity.* London, Free Press.

Schonberger, Richard J. 1986. *World Class Manufacturing: The Lessons of Simplicity Applied.* New York: The Free Press.

Schroeder, Roger G. 1993. *Operations Management: Decision Making in the Operations Function 4th ed.* New York: McGraw Hill.

Schumpeter, Joseph. 1975. *Capitalism, Socialism, and Democracy.* New York: Harper and Row, Chapter VII.

Siegel, Donald. 1997. "The Impact of Computers on Manufacturing Productivity Growth: A Multiple-Indicators, Multiple-Causes Approach." *Review of Economics and Statistics* 79(1): 68–78.

Siegel, Donald, David Waldman, and Albert Link. 1999. *Assessing the Impact of Organizational Practices on the Productivity of University Technology Transfer Offices: An Exploratory Study.* NBER working paper no. 7256, National Bureau of Economic Research, Cambridge, Massachusetts, July.

Skinner, Wickham. 1974. "The Focused Factory." *Harvard Business Review* 52(3): 113–121.

Snell, Scott A., and James W. Dean. 1992. "Integrated Manufacturing and Human Resource Management: A Human Capital Perspective." *Academy of Management Journal* 35(3): 457–504.

Spearman, C.S. 1904. "General Intelligence Objectively Determined and Measured." *American Journal of Psychology* 15: 201–293.

Thomas, Kenneth W., and Betty A. Velthouse. 1990. "Cognitive Elements of Empowerment: An Interpretive Model of Intrinsic Task Motivation." *Academy of Management Review* 15: 666–681.

Troske, Kenneth R. 1994. *Evidence on the Employer Size-Wage Premium from Worker-Establishment Matched Data.* Center for Economic Studies working paper 94-10, U.S. Census Bureau, April.

U.S. Bureau of the Census. (1989). *Current Industrial Reports-Manufacturing Technology 1988,* prepared under the direction of Gaylord Worden, U.S. Government Printing Office, Washington, D.C.

Van Reenen, John. 1996. "The Creation and Capture of Rents: Wages and Innovation in a Panel of U.K. Companies." *Quarterly Journal of Economics* 111: 195–226.

Van Reenen, John. 1997. "Employment and Technological Innovation: Evidence from U.K. Manufacturing Firms." *Journal of Labor Economics* 15(2): 255–284.

Welch, Finis. 1970. "Education in Production." *Journal of Political Economy* 78: 35–59.

Wendel, F.C., and R.G. Joekel. 1991. *Restructuring Personnel Selection: The Assessment Center Method.* Bloomington, Indiana: Phi Delta Kappa Educational Foundation.

Wolf, Gerrit, Manuel London, Jeff Casey, and Joseph Pufahl. 1995. "Career Experience and Motivation as Predictors of Training Behaviors and Outcomes for Displaced Engineers." *Journal of Vocational Behavior* 47: 316–331.

Zygmont, Jeffrey. 1987. "Manufacturers Move toward Computer Integration." *High Technology* (February): 16–21.

Author Index

Italicized parenthetical numbers following a page locator refer to the number of citations that an author received on that page. An italic *t* indicates a *table*.

126

Subject Index

The *italic letters e* and *t* following a page locator indicate that subject information is within an *equation* or *table,* respectively, on that page.

AAS. *See* Automated assembly system
Administrative employees
  employment levels and AMTs,
    67*t*–68*t,* 70*t,* 75*t,* 77*t*
  employment levels of, as key data
    item, 26, 38
Advanced manufacturing technologies
  (AMTs)
  accounting and inventory controls
    with, 89–90, 92, 96
  barriers to, 92–93, 95, 98
  benefits of, 83–84, 87, 93
  case studies of, 6, 81–96
  CIMS survey of Long Island firms, 4,
    30, 32*t*–33*t,* 109–114
  classification of, 6–9, 45–46, 58
  dynamic estimates of adoption, 63,
    65*t*
  government support for, 2–3, 92, 102
  implementation of, 5, 25–26, 31*t*–
    35*t,* 33–34, 45–46, 87–88, 100
  hypothesis on, 38, 41, 53, 55, 57
  list of specific, 72*t*
  probit estimates of adoption, 62–63,
    64*t*
  as proxy for technological change, 12
  workplace transformation with, 2, 94,
    99
  See also Integrated AMTs; Linked
    AMTs; Long Island, New York,
    survey
Aggregated data. *See* Data analysis
AGV. *See* Automated guided vehicle
AIL Systems (firm), defense contractor,
  74
American Standard (firm), ROB use, 52

AMTs. *See* Advanced manufacturing
  technologies
Annual Survey of Manufactures, 21
Arizona State University, 102–103
ASR. *See* Automated storage and
  retrieval
Automated assembly system (AAS), 71,
  72*t*
  difficulty in quantifying benefits of,
    93
Automated guided vehicle (AGV),
  51–52, 71, 72*t*
Automated storage and retrieval (ASR),
  51–52, 71, 72*t,* 83

B.F. Goodrich (firm), CAD systems in,
  47
Bar-code scanners, 47, 81–82, 89–90
Barriers
  to AMTs, 6, 9, 92–93, 95
  overcoming, 96, 105
Bayh-Dole legislation, 102
Bean (L.L.) company. *See* L.L. Bean
  (firm)
Benetton (firm), ASR systems, 51
BIC (firm), SPC, 49
Boeing (firm), 48, 74, 91

CAD. *See* Computer-aided design
CAE. *See* Computer-aided engineering
CAM. *See* Computer-aided
  manufacturing
Canada, technological change studies,
  15*t,* 17*t*
Career opportunities, as empowerment,
  1–2, 8, 54, 84, 99
Caterpillar (firm), JIT methods in, 50

About the Institute

The W.E. Upjohn Institute for Employment Research is a nonprofit research organization devoted to finding and promoting solutions to employment-related problems at the national, state, and local levels. It is an activity of the W.E. Upjohn Unemployment Trustee Corporation, which was established in 1932 to administer a fund set aside by the late Dr. W.E. Upjohn, founder of The Upjohn Company, to seek ways to counteract the loss of employment income during economic downturns.

The Institute is funded largely by income from the W.E. Upjohn Unemployment Trust, supplemented by outside grants, contracts, and sales of publications. Activities of the Institute comprise the following elements: 1) a research program conducted by a resident staff of professional social scientists; 2) a competitive grant program, which expands and complements the internal research program by providing financial support to researchers outside the Institute; 3) a publications program, which provides the major vehicle for disseminating the research of staff and grantees, as well as other selected works in the field; and 4) an Employment Management Services division, which manages most of the publicly funded employment and training programs in the local area.

The broad objectives of the Institute's research, grant, and publication programs are to 1) promote scholarship and experimentation on issues of public and private employment and unemployment policy, and 2) make knowledge and scholarship relevant and useful to policymakers in their pursuit of solutions to employment and unemployment problems.

Current areas of concentration for these programs include causes, consequences, and measures to alleviate unemployment; social insurance and income maintenance programs; compensation; workforce quality; work arrangements; family labor issues; labor-management relations; and regional economic development and local labor markets.